Proceedings of the Sixth Summer Institute in Theoretical Physics

SYMMETRY VIOLATIONS
IN SUBATOMIC PHYSICS

Proceedings of the Sixth Summer Institute in Theoretical Physics

SYMMETRY VIOLATIONS
IN SUBATOMIC PHYSICS

Kingston, Canada
18 – 29 July 1988

Editors

B. Castel
P.J. O'Donnell

World Scientific
Singapore • New Jersey • London • Hong Kong

Published by

World Scientific Publishing Co. Pte. Ltd.
P O Box 128, Farrer Road, Singapore 9128

USA office: World Scientific Publishing Co., Inc.
687 Hartwell Street, Teaneck, NJ 07666, USA

UK office: World Scientific Publishing Co. Pte. Ltd.
73 Lynton Mead, Totteridge, London N20 8DH, England

ISBN 9971-50-908-3

Printed in Singapore by JBW Printers & Binders Pte. Ltd.

PREFACE

These are the Proceedings of the Summer Institute in Theoretical Physics, the Sixth in the annual series of Institutes held every summer at Queen's University. The topic of the 1988 Institute was broadly defined as that of *the role of symmetries in sub-atomic physics*. The meeting brought together some of the most eminent experts in a wide range of physics who nevertheless were able to have very useful interactions within the four main topics; chiral symmetry, phase transitions, quantum chromodynamics (QCD) and eletroweak interactions.

The unifying role that symmetry brings to physics is evident by the variety of contributions given in these Proceedings. These range from the masslessness of the neutrino to phase transitions in the early universe: from chiral symmetry as an effective theory of QCD to a description of the properties of the nucleon; QCD theory on the lattice and as applied to deep inelastic scattering; heavy flavour rare decays and nuclear beta decays in the standard electroweak model.

The Summer Institutes in Theoretical Physics held at Queen's University on the shores of Lake Ontario are known as rather informal gatherings bringing together theoretical and experimental physicists for in-depth study seminars of usually two weeks duration. They are supported by the Natural Science and Engineering Research Council of Canada and Queen's University to whom the editors wish to express their sincere gratitude.

B. Castel and P. J. O'Donnell

CONTENTS

QCD — Reactions and Deep Inelastic Processes

Electroweak Interactions

Chiral Symmetry

ON THE ORIGIN OF CHIRAL LAGRANGIANS

John F. Donoghue
Department of Physics and Astronomy
University of Massachusetts
Amherst, MA 01003

ABSTRACT

I critically review some work on attempts to obtain the effective chiral Lagrangians from the underlying theory of QCD. The most promising avenue in my opinion, appears to be a connection of the chiral coefficients with the low-lying spectrum of the theory.

One of the goals of particle and nuclear physics is to "solve" QCD. For many, this means constructing heuristic models (quark models, pole models ...) which may have some ability to reproduce known data. This approach has some drawbacks as these models are inherently fuzzy. They do not really follow from QCD in any known approximation scheme, and adding some new features to the model may in fact make the model worse. They are not well controlled. The theory of chiral symmetry is different. It forms a rigorous and subtle approach to low energy QCD, handled as an expansion in the energy. The purpose of this talk is to describe some of the recent explorations by my collaborators (G. Valencia and C. Ramirez) and me, into the connection of effective chiral Lagrangians to QCD.

The language for describing chiral symmetry uses effective chiral Lagrangians[1,2,3]. Here for simplicity I will use only the massless limit in formulas, but results quoted from the literature will be for the true massive theory. The Lagrangians are organized in an expansion in the number of derivatives, i.e. in the energy.

$$L = \frac{F^2}{4} \, \mathrm{Tr}(\partial_\mu U \, \partial^\mu U^+) + \frac{\alpha_1}{4} \, \mathrm{Tr}(\partial_\mu U \, \partial^\mu U^+) \, \mathrm{Tr}(\partial_\nu U \, \partial^\nu U^+)$$

$$+ \frac{\alpha_2}{4} \, \mathrm{Tr}(\partial_\mu U \, \partial_\nu U^+) \, \mathrm{Tr}(\partial^\mu U \, \partial^\nu U^+) + \ldots$$

$$U = \exp \frac{i \, \tau \cdot \pi}{F} \tag{1}$$

with π^i being the pion field and $F = F_\pi = 93$ MeV the pion decay constant (the later relation is renormalized by loop effects). Phenomenologically we know the coefficients[2,4] α_1, α_2 etc. Contained in these parameters is information about the underlying theory, presumably QCD. What determines these?

The dominant answer in the literature[5] is that these coefficients arise in a relatively simple manner tied to the coefficient of the anomaly when one calculates the fermion determinant of QCD. Historically this grew out of attempts to understand the Wess Zumino Witten anomaly Lagrangian[6]. There are a variety of methods, with some common elements, applied in this study. In one typical example, one considers QCD coupled to an external chiral field.

$$L = -\frac{1}{4} \, F^2 + \bar{\psi} \, \slashed{D} \, \psi$$

$$\slashed{D} = \slashed{\partial} + i \, g \, A + \frac{1}{2} \, \partial_\mu U \, U^+ (1 + \gamma_5) + \frac{1}{2} \, U^+ \partial_\mu U \, (1 - \gamma_5) \tag{2}$$

such that the path integral over the fermion yields

$$\int d\psi \, d\bar{\psi} \exp i \, S = \exp \mathrm{Tr} \ln D = \exp \mathrm{Tr}\left[\ln D_o + \ln \frac{D}{D_o} \right] \tag{3}$$

where D_0 is the derivative with $U = 1$. One then expands $\mathrm{Tr} \ln \slashed{D}$ to some order, using heat kernel or zeta function methods. Technically one stops at the $\slashed{a}_2$ coefficient. Schematically,

$$\mathrm{Tr} \ln \frac{D}{D_o} = \int d^4x \, [\Lambda^4 + \Lambda^2 \, \mathrm{Tr}(\partial_\mu U \, \partial^\mu U^+)]$$

$$- \frac{Nc}{384\pi^2} \, [[\mathrm{Tr}(\partial_\mu U \, \partial^\mu U^+)]^2 - 2[\mathrm{Tr}(\partial_\mu U \, \partial_\nu U^+)]^2 + \ldots] + Nc \, \Gamma wzw, \tag{4}$$

with Λ being a cutoff and with Γwzw being the anomaly effective Lagrangian. At this order then α_1 and α_2 appear at the same time as the anomaly. The result is

$$\alpha_1 = \frac{-1}{32\pi^2} \approx -0.003$$

$$\frac{\alpha_2}{\alpha_1} = -2 \tag{5}$$

Although at first sight this appears impressive, I am now convinced that it is wrong. What are the problems? First, it does not agree with experiment (contrary to claims in the literature). A tree level fit to the data[4] (the appropriate comparison) gives
$$\alpha_1 = -0.0092$$

$$\frac{\alpha_2}{\alpha_1} = -0.85 \tag{6}$$

Comparison with the running renormalized parameter of Gasser and Leutwyler[2] is inappropriate as their parameters have an infinity absorbed into them, have a scale dependence and amplitudes using them have finite loop effects remaining which depend on the renormalization prescription. These features also show that a fixed number cannot be the full answer for a renormalized coefficient. Another problem is that the calculation is independent of the gluons in QCD. It is simply the free quark loop. This is surely unrealistic. In the case of the anomaly, the Adler Bardeen theorem[7] guarantee that the free quark loop is not modified by gluonic corrections. No such result applies to any other term. Technically the flaw comes from stopping at a_2 in the expansion of the fermion determinant. These coefficients are dimensionless and QCD has no small mass ratio that can be used as an expansion parameter. All terms in the expansion can contribute. For an example a higher order term could have the operator form

$$F^A_{\mu\nu} F^{A\mu\nu} [\mathrm{Tr}(\partial_\mu U \partial^\mu U^+)]^2 \tag{7}$$

This is missed by stopping at the a_2 coefficient. However when integrated over the gluon field it will also produce a contribution to α_1. Pictorially the results can be given as in Fig. 1 where the x indicates the action of an external chiral field. The free result is given in Fig. 1a, b, while gluonic modifications are fin Fig. c, d. It is the sum of all of these which yields the effective Lagrangian.

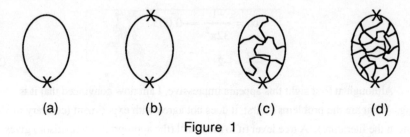

$$(a) \qquad (b) \qquad (c) \qquad (d)$$

Figure 1

This analysis also suggests physics which might be important in a realistic determination of the coefficients. Consider Fig. 1a, with all possible gluonic effects. If the x were to represent a quark operator with a given quantum number, this figure could represent the way that hadron masses are measured in quenched lattice QCD. There the loop is equivalent to a sum over meson propagators as in Fig. 1b, with the long distance correlation yielding the mass of the lightest state. However a combination of chiral fields coupled to fermion could be used equally well to create the $q\bar{q}$ loop. This is then just the loop calculation that appears in the fermion determinant, which could then also presumably be written as a sum over particle propagator. This suggests a connection of the parameter in a chiral Lagrangian with the spectrum of the theory[8].

This connection can be made more precise by considering $\pi\pi$ scattering, which is governed by the chiral Lagrangian. The lowest order prediction (in the energy) is fixed entirely in terms of F_π, and carries no extra information besides the value of F_π. This is shown vs. The data as the dashed line in Fig. 2, in the $I = 1, J = 1$ channels. [There are 5 channels where this can be applied. The reader is referred to Ref. 4 for a more detailed description.] the next order Lagrangian, however, is sensitive to the presence of the resonance, the ρ in this case. This connection is required by unitarity. These corrections increase the prediction to match the tail of the ρ. In the linear σ model, without a ρ, the correction would go in the opposite direction. In order to match the tail of the ρ in the s and t channels would require

$$\alpha_1 = 48\pi \, \frac{\Gamma_\rho F_\pi^4}{m_\rho^5} = -0.0084$$

$$\frac{\alpha_1}{\alpha_2} = -1 \tag{8}$$

in very close agreement with the experimental numbers.

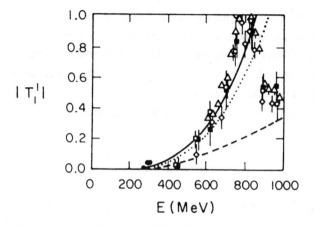

$|T_I^I|$

E (MeV)

Figure 2

My colleagues and I have been exploring this idea in more general contexts[8]. We find that in some simple solvable but nontrivial models that the low-lying spectrum is the prime determinant for the E^4 coefficients in the Lagrangian. One example is the linear σ model treated in the perturbative regime. There the light resonance in the $I = 0, J = 0$ channel determines all the physics in the Lagrangian. Another case is the so called gauged chiral model where vector and axial vector fields are added with Young Mills like self coupling. In these cases the presence of other interactions does not significantly modify the effect of the spectrum. We have developed criteria for when a particular resonance is light enough to have an influence. In all cases coupled to $\pi\pi$, the effect scales as $\dfrac{\Gamma_i F_\pi^4}{m_i^5}$, where Γ is the width, so that an increase in mass quickly decreases the contribution. We have also suggested operational methods for obtaining the coefficients when working to one loop order. As noted above, this seems to work well phenomenologically in the real world.

8

We have hopes of applying these ideas in other areas. For example in QCD, one could use the resonance connection to sum a set of diagrams to all orders in the energy expansion by keeping the ρ propagator in the Feynman diagrams of ρ exchange. This is a connection with pole models and vector dominance, but in the more controlled setting of chiral symmetry. In a different area these ideas could be used to explore physics beyond the standard model (such as Technicolor) which also use effective Lagrangians.

REFERENCES

1] S. Weinberg, Physica 96A, 327 (1979).

2] J. Gasser and H. Leutwyler, Ann. Phys. 158, 142 (1984); Nucl. Phys. B250, 465 (1985).

3] J.F. Donoghue in Proc. III Int. Conf. on Intersections of Particle and Nuclear Physics, Rockport, Maine, ed. by G. Bunce, to be published.

4] J.F. Donoghue, C. Ramirez and G. Valencia, Phys. Rev. D38, 360 (1988).

5] J. Balog, Phys. Lett. 149B, Nucl. Phys. B258, 361 (1985).

   A. Andrianov, Phys. Lett. 157B, 425 (1985).

   H.Y. Cheng, Phys. Rev. D34, 166 (1981).

   L.H. Chan, Phys. Rev. Lett. 55, 21 (1985).

6] J. Wess and B. Zumino, Phys. Lett. 37B, 95 (1971).

   E. Witten, Nucl. Phys. B223, 422, 433 (1983).

7] S. Adler, Phys. Rev. 177, 2426 (1969)

   W.A. Bardeen, Phys. Rev. 184, 1848(1969).

8] J.F. Donoghue, C. Ramirez and G. Valencia, UMHEP-298.

The Chiral Bag Model with Vector Mesons

Hiroshi Toki*

Department of Physics, Queen's University, Kingston K7L 3N6, Canada

1. Introduction and QCD

What is the nucleon? This was the question many nuclear physicists started to ask, when they heard often the word 'swollen nucleon' and the discussion of the future machine to see 'quarks' in nuclei. Although the final answer is far from being reached, I would like to present here the reasoning of my interest in studying the chiral bag model. Since we have a lot of free time in this Summer Institute, the details will be discussed privately.

To start with, let us see what quantum chromodynamics (QCD) tells us:

1. Asymptotic freedom: The deep inelastic scattering experiment reveals the asymptotic free nature of quarks.

2. Quark and gluon confinement: The search of a free quark has failed up to now.

3. Chiral symmetry: The current algebra is successful in explaining many low energy hadron properties.

The QCD Lagrangian has the faimiliar form:

$$L_{QCD} = \bar{\psi}(i\not{D} - m)\psi - \frac{1}{4}F_{\mu\nu}F^{\mu\nu}, \tag{1}$$

*Permanent address; Department of Physics, Tokyo Metropolitan University, Setagaya, Tokyo 158, Japan where quarks ψ and gluons are the dynamical degrees of freedom. The non-abelian nature of the field equation makes the attempt to solve the QCD equations prohibitively difficult. The numerical method (lattice QCD) is being used but is still very far from providing a clear picture of the nucleon. Hence, it is very important to construct a phenomenological model for the nucleon. In order to have some contact with QCD, any model would have to take into account the above three properties of QCD. We may then hope to justify the phenomenological model by QCD.

2. <u>Phenomenological Models</u>

Let us examine the phenomenological models in the literature. We start with the MIT bag model[1]. The Lagrangian is

$$L_{MIT} = (\bar{\psi}i\delta\psi - B)\theta_B + \bar{\psi}\psi\delta_B \qquad (2)$$

The quark field ψ is confined with a bag, which is indicated by $\theta_B = \theta(R - r)$ with R the bag radius. The difference between the normal vaccum (Wigner mode) inside and the nontrivial vacuum (Goldstone mode) outside the bag appears in this model as the bag pressure B, i.e. a phenomenological parameter. The major gluon contributions are supposed to be contained in B and only the one gluon exchange interaction contribution is added perturbatively. (This part is not shown in Eq. 1). The boundary condition of the quark field is given by the second term with $\delta_B = \delta(r - R)$. The MIT bag model is extremely successful in explaining the low-lying hadron properties.

Although successful, there are a few unwanted features:

1. $\alpha_g \sim 2$: The gluon-quark coupling constant α_g is too big for the perturbation method to be justified.

2. $R_{\text{bag}} \sim 1\ fm$: The size of the nucleon comes out to be too large. There is no

room left for meson exchange currents to play a role in nuclear physics. The area of quarks may be too large to utilize the asymptotic freedom idea.

3. No chiral invariance: The chiral invariance is not incorporated in the model. This lack of the important symmetry gave a motivation to introduce pions outside the bag, the chiral bag[2], to which I shall come back later.

The model we would like to discuss next is the completely opposite picture for nucleon. The Skyrme model was invented by Skyrme long ago[3], in which a nucleon is identified as a topological soliton of a chiral Lagrangian;

$$L_{SK} = \frac{f_\pi^2}{4} tr(\partial_\mu U \partial^\mu U^+) + \frac{1}{32e^2} tr[U \partial_\mu U^+, U \partial_\nu U^+]^2, \tag{3}$$

where U is the SU(2) chiral field, $U = e^{i \vec{\tau} \cdot \vec{\pi}/f_\pi}$. f_π is the pion decay constant and e the Skyrme parameter. This model respects the chiral invariance, which is realized in the Goldstone mode. The biggest support of this model comes from the analysis of the QCD at large N_c (color number) limit, which suggests that the low energy behaviour of QCD is expressed in terms of weakly interacting meson fields[4]. Quantitatively this model reproduces static properties of the nucleon within about 30% ($\sim 1/N_c$)[5].

This model satisfies all the three features of QCD. My personal complaint is the absence of quarks and gluons. Where are these fields of QCD? How does the parton picture come about. From a practical point, how can we improve on this model?

3. Toward Improvement

There are three directions to choose. The first is to add higher derivative terms; the line of Skyrme[3]. Gasser and Leutwyler have studied the meson properties with the

Lagrangian[6]);

$$L_{GL} = c^{(2)}tr[L_\mu L^\mu] + c_1^{(4)}tr[L_\mu, L_\nu]^2 + c_2^{(4)}tr\{L_\mu, L_\nu\}^2 + c^{(6)}\{...\} + ..., \qquad (4)$$

where $L_\mu = U^+\partial\mu U$. The dynamical degree of freedom is only the chiral SU(2) field U, which is much too restrictive. As a result, the magnetic form factor of the nucleon is related with the electric form factor[5]. Meissner et al.[7], showed that the isovector magnetic moment is related with the nucleon and delta masses; $\mu_p - \mu_n = M_N/(M_\Delta - M_N)$, whatever the values we use for c's in L_{GL}. This relation is badly broken experimentally (32%).

The second attempt is to add more mesons; the line of Witten[4]. The least heavy mesons after pion in the non-strange sector are the vector mesons. Vector mesons are also known to play an important role for electromagnetic properties (vector dominance[8]) and for the nucleon-nucleon interaction. There are two systematic ways to introduce vector mesons. In the massive Yang Mills scheme (MYM), the SU(2) × SU(2) symmetry is gauged and the gauge bosons are identified to vector V_μ and axial vector A_μ mesons[9];

$$L_{MYM} = \frac{f_\pi^2}{2}tr(D_\mu U D^\mu U^+) - 2f_\pi^2 tr(V_\mu^2 + A_\mu^2) + \frac{1}{2g^2}tr(V_{\mu\nu}^2 + A_{\mu\nu}^2) \qquad (5)$$

with the covariant derivative

$$D_\mu U = \partial\mu U + [V_\mu, U] - \{A_\mu, U\}$$

and g is the gauge coupling constant. The hidden local symmetry scheme (HLS) was found recently by Bando et al., who splitted the chiral SU(2) field into two pieces $U = \xi_L^+\xi_R$ and introduced vector and axial vector mesons as the gauge bosons of the hidden local symmetry[10]. The resulting Lagrangian is

$$L_{HLS} = -2f_\pi^2 tr[\tilde{V}_\mu - \frac{1}{2}(\xi_R\partial\mu\xi_R^+ + \xi_L\partial\mu\xi_L^+)]^2$$
$$-2f_\pi^2 tr[\tilde{A}_\mu - \frac{1}{2}(\xi_R\partial\mu\xi_R^+ - \xi_L\partial\mu\xi_L^+)]^2 \qquad (6)$$
$$-2f_\pi^2 tr\tilde{A}_\mu^2 + \frac{1}{2g^2}tr(\tilde{V}_{\mu\nu}^2 + \tilde{A}_{\mu\nu}^2)$$

These two Lagrangians are found equivalent for all the meson dynamics[11,12], which is shown explicitly by using the Stückelberg transformation[13];

$$\tilde{V}_\mu + \tilde{A}_\mu = \xi_R(V_\mu + A_\mu + \partial\mu)\xi_R^+$$

$$\tilde{V}_\mu - \tilde{A}_\mu = \xi_L(V_\mu - A_\mu + \partial\mu)\xi_L^+ \qquad (7)$$

We remark also that Wakamatsu and Weise showed that the above forms are derived from the extended Nambu-Jona-Lasinio Lagrangian by bosonizing the quark bilinears[14]. These Lagrangians have remarkable properties; the KSFR relation

$$m_V^2 = 2f_\pi^2 g^2$$

and the Weinberg mass relation

$$m_A^2 = 2m_V^2$$

In addition, the vector dominance properties are found by introducing the electromagnetic fields[10].

Using these Lagrangians with ω meson included, nucleon is constructed by various authors[14,12,7]. In general they found good nucleon properties except the nucleon mass M_N and the axial coupling constant g_A. We believe the reason for the discrepancies stems on the spin-isospin projection. We ought to use a better projection technique like the generator coordinate method (GCM), which is demonstrated powerful in nuclear physics. The difference between the standard method and the GCM is demonstrated for the nucleon mass[16] and for $g_A^{17)}$.

Although successful, there is a problem inherent in this model which is the $\kappa_\rho - \kappa_v$ problem [18]. κ_ρ is the ratio of the tensor coupling against the vector coupling of the meson with nucleon. Experimentally, this ratio is $\kappa_\rho = 6.6 \pm 0.6$. κ_v is the similar ratio for the photon-nucleon coupling which is $\kappa_v = 3.7$ experimentally. Since the vector dominance implies that photon couples with nucleon through vector mesons, these ratios should be equal. The actual calculation shows that the ratios are $\kappa_\rho = 5.38$ and $\kappa_v = 4.93$[19], the small difference of which is caused from the higher order terms[7]. Brown, Rho and Weise[18] claim that this is the place quarks show up, since photon can couple with quarks directly. This direct coupling of photon with a component of nucleon breaks the near equality between κ_ρ and κ_v. This would be a strong motivation to look into the (hybrid) chiral bag model, where quarks are confined in a bag and a partial Skyrmion surrounds the quark bag.

4. Chiral bag model and the Chesire cat principle

This discussion brings us naturally to the third model, which includes quarks but still pions are the main ingredients. The chiral bag model Lagrangian is[20,21]

$$L_{CB} = (\bar{\psi} i \delta \psi - B)\theta_B + L_{SK}\theta_{\bar{B}} + \bar{\psi}e^{i\gamma_5 \vec{\tau}.\vec{\pi}/f_\pi}\psi\delta_B \tag{8}$$

Quarks are confined in a bag, while pions are excluded from the bag; L_{SK} is the Skyrmion Lagrangian (3). The third term is chiral invariant and relates quarks and pions at the bag boundary, $\delta_B = \delta(r - R)$.

At this place we have to discuss the Chesire cat principle. This principle says that the presence of the bag does not have a physical meaning; i.e. all the physical quantities do not depend on the bag radius[22]. One can verify it in the $1 + 1$ dimensional case[23], but it is merely a belief for the $1 + 3$ dimensional case. This principle may or may not be true. But to be true, the model has to be complete at any bag radius. Let us imagine how the nucleon is to be described. In the limit $R \to 0$, the dynamical fields are mesons. As discussed above, pions and vector mesons are not enough. We would need a large amount of meson fields to describe nucleon; $|N> = \sum_{i=1}^{\infty} a_i |M_i>$. In the other limit $R \to \infty$, the dynamical fields are quarks and gluons. The asymptotic freedom is not applicable and one ought to deal with many quarks-antiquarks excitations with gluons; $|N> = \sum_{\substack{i=1 \\ j=1}}^{\infty} b_{ij} |q^3; (q\bar{q})^i, g^j>$. At a reasonable size, we hope that the heavy mesons become inactive and the asyumptotic freedom feature still remains and therefore nucleon is describable with few terms; $|N> = \sum_{i=1}^{\text{few}} a_i |M_i> + b|q^3>$. We believe this is only the case which we can work out. In addition, it is often so that the right picture is the most economical one. We therefore think that the chiral bag model is applicable only for reasonable bag sizes ($R \sim 0.5 fm$).

Many authors have contributed to the chiral bag model[24]. The Casimir contributions to all the quantities have to be calculated. The results are often divergent. We ought to develope a method to treat these divergences. At this moment, the divergence was thrown out by hand with the constraint that the chiral bag model reduces to the Skyrmion at $R \to 0$. There seems a slight improvement but the results are qualitatively unchanged from the ones of the Skyrmion. Hence, the necessity of quarks is not convincing.

5. Chiral bag model with vector mesons

These discussions naturally lead to an idea to construct the chiral bag model with vector mesons. The Lagrangian

$$L_{CBV} = (\bar{\psi}i\slashed{\partial}\psi - B)\theta_B + L_{MYM}\theta_{\bar{B}} + \bar{\psi}e^{i\gamma_5\vec{\tau}\cdot\vec{\pi}/f_\pi}\psi\delta_B \tag{9}$$

Each term has been discussed up to now. This model is attractive, since it contains all the phenomenological aspects and the QCD requirments. Quarks are confined in a small space. The gluon and the quark condensation effects are summarized in the bag constant B. The mesons, the active excitation modes in the Goldstone phase, are excluded from the gag. These mesons have the vector dominance properties. The chiral invariance makes the bridge between the inside and the outside of the bag. We are pretty close to getting the final results for nucleon properties.

Since there was a confusion, we would like to discuss the procedure in getting the hedgehog solution of the Skyrmion with the vector mesons. The aim here is to show the equivalence between the MYM and the HLS schemes for the Skyrmion properties[25]. The Lagrangians have been given in Eqs. 5 and 6. In order to construct the soliton with the baryon number $B = 1$, we introduce the hedgehog ansatz;

$$
\begin{aligned}
U &= \exp[i\vec{\tau}\hat{r}F(r)] \\
V_i^a &= \varepsilon_{aij}\hat{r}_j\frac{G(r)}{r} \\
A_i^a &= A(r)P_{ai} + B(r)\hat{r}_a\hat{r}_i \\
\omega_0 &= \omega(r)
\end{aligned}
\tag{10}
$$

with $\omega_i = V_0^a = A_0^a = 0$. $P_{ai} = \delta_{ai} - \hat{r}_a\hat{r}_i$, which is orthoganal to $\hat{r}_a$ and $\hat{r}_i$. The same expressions are used for the HLS scheme, where we introduce the notation $\tilde{G}, \tilde{A}$ and

$\tilde{B}$, respectively. The soliton energies are found in the two schemes;

$$
\begin{aligned}
E_{MYM} = 4\pi \int r^2 dr[& \frac{1}{g^2}(AB - \frac{G'}{r})^2 + \frac{1}{2g^2}(A^2 + \frac{G(G-2)}{r^2})^2 \\
& + \frac{1}{g^2}(A' + \frac{A-B}{r} + \frac{BG}{r})^2 + \frac{a}{2}f_\pi^2(\frac{G}{r})^2 + 2A^2 + B^2) \\
& + \frac{a}{a-1}\frac{f_\pi^2}{2}((B+F')^2 + 2(A\cos F + \frac{1-G}{r}\sin F)^2) \\
& - \frac{1}{2}\omega'^2 - \frac{1}{2}g^2 f_\pi^2 \omega^2 + \frac{N_c}{2}g\omega\frac{\sin^2 F}{2\pi^2 r^2}F'],
\end{aligned}
\tag{11}
$$

$$
\begin{aligned}
E_{HLS} = 4\pi \int r^2 dr[& \frac{1}{g^2}(\tilde{A}\tilde{B} - \frac{\tilde{G}'}{r})^2 + \frac{1}{2g^2}(\tilde{A}^2 + \frac{\tilde{G}(\tilde{G}-2)}{r^2})^2 \\
& + \frac{1}{g^2}(\tilde{A}' + \frac{\tilde{A}-\tilde{B}}{r} + \frac{\tilde{B}\tilde{G}}{r})^2 + af_\pi^2(\frac{\tilde{G}}{r} - \frac{1-\cos F}{r})^2 \\
& + af_\pi^2((\tilde{A} - \frac{\sin F}{r})^2 + \frac{1}{2}(\tilde{B} - F')^2) + \frac{a}{a-1}f_\pi^2(\tilde{A}^2 + \frac{1}{2}\tilde{B}^2) \\
& - \frac{1}{2}\omega'^2 - \frac{a}{2}g^2 f_\pi^2 \omega^2 + \frac{N_c}{2}g\omega\frac{\sin^2 F}{2\pi^2 r^2}F'],
\end{aligned}
\tag{12}
$$

The Stückelberg transformation (7) relates the hedghog profiles in the two schemes;

$$
\begin{aligned}
\tilde{G} &= G + rA\sin F + (1-G)(1-\cos F) \\
\tilde{A} &= A\cos F + \frac{1-G}{r}\sin F \\
\tilde{B} &= B + F'
\end{aligned}
\tag{13}
$$

We can show explicitly by inserting the above relations (13) to the energy (12) that the two energy expressions transform to each other. Since the profile functions are related linearly[13], these two schemes provide identical results for nucleon properties.

We show in Table 1 the numerical results on the hedgehog soliton properties with the mesonic parameters taken at their physical values, $f_\pi = 93$ MeV, $g = 5.85$. It is interesting to point out the role of the A_1 meson on the soliton properties. The hedgehog mass M_H is reduced largely from 1480 MeV to 1200 MeV, while the axial coupling constant g_A is found smaller by inclusion of the A_1 meson. The soliton size, as seen from the radius of the hedgehog soliton r_H and the isoscalar radius $r_{I=0}$, does not change much.

The similar equivalence holds also for the case of the chiral bag model with vector mesons. In addition, the equivalence still is found to persist even after performing the semi-classical quantization.

We would like to show here a few results on the chiral bag model with vector mesons. The parameters in the Lagrangian (9) are completely fixed by the meson properties. Shown in Fig. 1 is the behaviour of the hedgehog energy as a function of the bag radius R. At $R = 0$, the hedgehog energy of $B = 1$ is dominated by the meson vector (Skyrmion). As R increases, the meson contribution decreases and the quark contribution increases. At $R \sim 0.5 fm$, the two contributions are about equal. The behaviours of the axial coupling constant g_A of the nucleon and the baryon radius are shown in Fig. 2. g_A drops first and then increases up to $R \sim 1 fm$. We believe the better spin-isospin projection would enhance g_A about 5/3 and the curve with 5/3 multiplied is also shown. The experimental value 1.25 seems to be met ar $R \sim 0.5 fm$. An interesting result is observed in the lower part of Fig. 2. The baryon radius R_B increases rather monotonically with R. On the other hand, the isoscalar nucleon radius $R_I = 0$, which is observed by photon is almost constant up to the bag radius $R \sim 1 fm$. This behaviour comes from the decreasing role of ω-meson as R increases. Other quantities need spin-isospin projections, which are nearly completed.

Acknowledgement

The author acknowledges fruitful collaborations with Atsushi Hosaka and Wolfram Weise on the chiral bag model with vector mesons. He is grateful to Boris Castel for his support and hospitality during his stay at Queen's University.

20

References

1. A. Chodos, R.L. Jaffe, K. Johnson, C.B. Thorn and V.F. Weisskopf, Phys. Rev. D9 (1974) 3471.

   T.A. DeGrand, R.L. Jaffe, K. Johnston and J. Kiskis, Phys. Rev. D12 (1975) 2060.

2. A. Chodos and C.B. Thorn, Phys. Rev. D12 (1975) 2733.

   T. Inoue and T. Maskawa, Prog. Theor. Phys. 54 (1975) 1833.

3. T.H.R. Skyrme, Proc. Roy. Soc. London 260 (1961) 127.

   T.H.R. Skyrme, Nucl. Phys. 31 (1961) 556.

4. G. t'Hooft, Nucl. Phys. B72 (1974) 461.

   E. Witten, Nucl. Phys. B160 (1979) 57.

5. G.S. Adkins, C.R. Nappi and E. Witten, Nucl. Phys. B228 (1983) 552.

6. J. Gasser and H. Leutwyler, Phys. Lett. B184 (1987) 83.

7. U.-G. Meissner, N. Kaiser and W. Weise, Nucl. Phys. A466 (1987) 685.

8. J.J. Sakurai, Currents and Mesons (Chicago University Press, Chicago 1969).

9. O. Kaymakcalan, S. Rajeev and J. Schechter, Phys. Rev. D30 (1984) 594.

10. M. Bando, T. Kugo, S. Uehara, K. Yamawaki and T. Yanagida, Phys. Rev. Lett. 54 (1985) 1215.

11. K. Yamawaki, Phys. Rev. D35 (1987) 412.

12. U.-G. Meissner and I. Zahed, Z. Physik **A327** (1987) 5.

13. E.G.C. Stückelberg, Helv. Acta. Phys. **14** (1941) 51.

14. M. Wakamatsu and W. Weise, Z. Physik A, to be published (1988).

15. U.-G. Meissner and I. Zahed, Phys. Rev. Lett. **57** (1986) 170.

16. A. Hosaka, Phys. Lett., to be published.

    M.C. Birse, Phys. Rev. **D33** (1986) 1934.

17. A. Hosaka, K. Kusaka, H. Takashita and H. Toki, Prog. Theor. Phys. **76** (1986) 315.

18. G.E. Brown, M. Rho and W. Weise, Nucl. Phys. **A454** (1986) 669.

19. N. Kaiser, U. Vogel, W. Weise and U.-G. Meissner, Regensburg preprint, TPR-88-5 (1988).

20. M. Rho, A.S. Goldhaber and G.E. Brown, Phys. Rev. Lett. **51** (1983) 747

    G.E. Brown, A.D. Jackson, M. Rho and V. Vento, Phys. Lett. **B140** (1984) 285.

21. A. Hosaka and H. Toki, Phys. Lett. **B167** (1986) 153.

22. H.B. Nielsen, 'Skyrmions and Anomalies' World Scientific (1988)

23. S. Nadkarni, H.B. Nielsen and I. Zahed, Nucl. Phys. **B253** (1985) 308.

24. J. Goldstone and R.L. Jaffe, Phys. Rev. Lett. **51** (1983) 1518.

    I. Zahed, A. Wirzba and U.-G. Meissner, Ann. Phys. **165** (1985) 406.

    P.J. Mulders, Phys. Rev. **D30** (1984) 1073.

25. A. Hosaka, H.Toki and W. Weise, Z. Physik A, to be published (1988).

Table 1

Hedgehog Soliton Properties

	present result	result without A_1
M_H	1200 MeV	1480 MeV
r_H	0.47 fm	0.50 fm
$r_{I=1}$	0.78 fm	0.80 fm
g_A	0.70	0.88

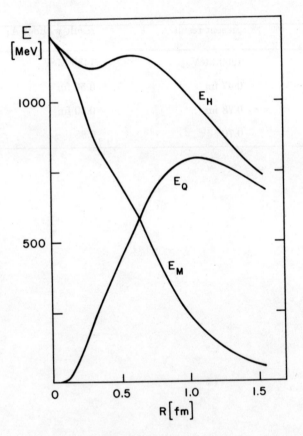

Fig. 1: The hedgehod energy E_H as a function of the bag radius R. E_Q is the quark contribution and E_M the meson contribution.

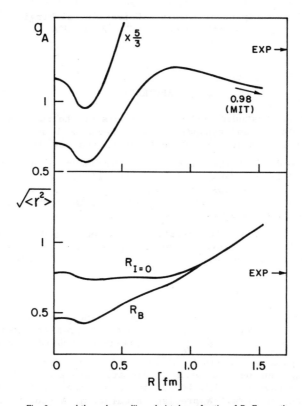

Fig. 2: g_A and the nucleon radii are depicted as a function of R. For g_A, the result
with 5/3 being multiplied is shown for R smaller than 0.5 fm to indicate that a better
spin-isospin projection method would result in this enhanced value.

Axial Charge Renormalization of Isolated
and Bound Nucleons

J.V. Noble[*]

Institute of Nuclear and Particle Physics
Department of Physics, University of Virginia
Charlottesville, VA 22901   USA

ABSTRACT

This paper studies structural models of the nucleon, with
the aim of understanding the renormalized axial charge
g_A, its relation to πN scattering, and how it might be
affected by the nuclear environment. The models
considered include the bag- and linear σ- and $\sigma + \omega$
models, since most hybrid models (chiral bag, chiral
soliton, Skyrmion, *etc.*) resemble some combination of
these. The fact that g_A is finite in a renormalized field
theory — *i.e.*, it satisfies an unsubtracted dispersion
relation — allows a cutoff-insensitive perturbation-
theoretic estimate.

We find the σ-model alone predicts too large a g_A even
with reasonable cutoffs; that ω meson renormalization
corrections are large and negative and correct the ex-
cessive σ- and π- loop contributions from the σ-model;
and that the "bare" g_A in quark-meson hybrid models is
uncertain by -20%+60%. Hence, while the observed axial
charge can be fit with parameters consistent with all
hadron physics below 1 GeV, there are too many uncertain
aspects to permit a precise calculation. A good value of
g_A is not a critical test of baryon model.

[*]Supported in part by the US National Science Foundation

1. INTRODUCTION

Structural models of the hadrons based on ideas from quantum chromodynamics (QCD) have in recent years come to resemble the old picture
of a compact "bare" nucleon surrounded by a meson cloud. The "bare"
nucleon is represented by a "little bag"[1] of valence quarks, and the
meson cloud by a Skyrmion or other such semiclassical field configuration. We would like these models to agree reasonably well with
known baryon properties: masses, magnetic moments, charge radii and
axial charges. Several authors seem to regard getting the axial charge
right as an especially crucial test of nucleon structure models.

Since the Adler-Weisberger sum rule[2] has been around for more than two
decades, it might seem somewhat pedantic to insist on a good value for
the axial charge of the nucleon. However, the sum rule and equivalent soft-pion relations between g_A and pion-nucleon scattering
amplitudes[3] tell us that to treat πN scattering properly our models
must — at least — agree with g_A, and *vice versa*.

The object of this note is to clarify the problem of calculating g_A in
various (related) models of baryon structure. The focus is on the
precision we may expect at our present level of understanding of the
underlying theory, as well as on the possible modifications of the
axial charge of a nucleon in nuclear matter.

2. NAIVE QUARK (BAG) MODEL

In the simplest $SU(3)_c$ quark model of the nucleons, the proton wave
function has the form

$$\left\{ 2u_\uparrow u_\uparrow d_\downarrow + 2u_\uparrow d_\downarrow u_\uparrow + 2d_\downarrow u_\uparrow u_\uparrow \right.$$
$$-u_\uparrow d_\uparrow u_\downarrow - u_\downarrow u_\uparrow d_\uparrow - u_\downarrow d_\uparrow u_\uparrow$$
$$\left. -d_\uparrow u_\uparrow u_\downarrow - d_\uparrow u_\downarrow u_\uparrow - u_\uparrow u_\downarrow d_\uparrow \right\} \Big/ \sqrt{18} \ . \tag{1}$$

That is, we represent the nucleons as properly symmetrized, color-zero product states of single-particle orbitals in a confining potential. The axial charge is defined to be ($d^\dagger(\vec{x})$ and $u(\vec{x})$ are quark field operators)

$$g_A = \int d^3x \, \langle n_\uparrow | d^\dagger(\vec{x}) \sigma_z u(\vec{x}) | p_\uparrow \rangle \ . \tag{2}$$

Thus, if the Dirac single particle wave function of a quark is

$$\psi(\vec{r}) = (4\pi)^{-1/2} \begin{pmatrix} a(r) \\ i\vec{\sigma}\cdot\hat{r}b(r) \end{pmatrix} \chi \ , \tag{3}$$

where $\psi(\vec{r})$ is normalized,

$$\int_0^\infty dr r^2 \left[a^2(r) + b^2(r) \right] = 1 \ , \tag{4}$$

the axial charge becomes

$$g_A = \frac{5}{3} \int_0^\infty dr r^2 \left[a^2(r) - \frac{1}{3}b^2(r) \right] \ . \tag{5}$$

The nonrelativistic limit of Eq. 5 is $\frac{5}{3}$; while with relativistic wave functions, g_A depends on the nature and form of the confining potential[4]. The fact that the nonrelativistic (upper) limit $g_A \leq 1.67$ exceeds the experimental value 1.25-1.28 has no significance, since as is well known[1], the simple bag model does not respect chirality conservation, hence cannot lead to a partially conserved axial current (PCAC). Thus we must expect important contributions to g_A from any additional physics incorporated to restore axial symmetry.

Moreover, there is good reason to believe the valence quarks are highly relativistic (as would be the case for light "current" quarks confined in a bag of radius ≤ 0.5 fm): the "small" Dirac components then cancel a significant fraction of the axial charge. For a rough idea how much cancellation occurs, suppose the neutron-proton mass difference arises from the difference of (current) quark masses plus electromagnetic field (mostly Coulomb) energy:

$$M_n - M_p \approx \left(m_d - m_u\right) \int_0^\infty dr r^2 \left[a^2(r) - b^2(r)\right] - \Delta E_{e.m.} \qquad (6)$$

Suppose $\Delta E_{e.m.} \approx 0.5\text{-}1.5$ MeV and $m_d - m_u \approx 3\text{-}7$ MeV[5]; then we can estimate the integral in Eq. 5 to be

$$\int_0^\infty dr r^2 \left[a^2(r) - \frac{1}{3}b^2(r)\right] \approx 0.50 \text{ to } 0.96 , \qquad (7)$$

giving

$$g_A \approx 0.8\text{-}1.6 . \qquad (8)$$

That is, the unknown aspects of QCD lead to a broad range of possible values of the axial charge, even before we take into account the polarization effects ("sea" quarks) that will surely modify g_A.

3. HYBRID MODELS OF BARYON STRUCTURE

Hybrid models of the nucleon are based on the evident fact that the active hadronic degrees of freedom at low excitation energy are baryons and mesons. Precisely how QCD manages to arrange matters to "freeze out" quark and gluon degrees of freedom is still conjectural[6]. Operationally, one conjectures both an effective meson lagrangian and a prescription for restoring chiral symmetry through quark-meson phenomenological couplings, then proceeds to calculate, usually by solving semiclassical equations of motion derived from a minimum principle. Typically, the hybrid models have fallen into two categories: Skyrmion models[7] and chiral soliton models[8]. The latter are rather similar — except for taking the Fermions of the theory to be quarks — to the linear σ-model of Gell-Mann and Levy[9]; whereas the former — aside from the higher- derivative Skyrme terms — resemble the (nonlinear) Weinberg model[10].

The similarities noted above lead us to study the renormalization of g_A in the linear σ-model, to 1-loop order. The linear σ-model respects chiral symmetry and the Adler consistency condition[11], and embodies spontaneously broken symmetry. Since it is renormalizable we may perturb about the vacuum value of the σ field, secure in the knowledge that our results will be consistent, if not necessarily complete. Moreover, approaching the problem from this point of view lets us include the quantum fluctuations necessarily omitted from the semiclassical calculations of hybrid models. We shall find that to obtain reasonable values for g_A we must incorporate the ω meson (in a renormalizable fashion[12]). That is, the ω is a necessary ingredient — either directly as part of a phenomenological theory of mesons and baryons, or virtually as a dominant part of the "sea" in QCD — of a theory of baryon structure. This has been noted previously in the context of chiral soliton models[13].

4. THE ADLER-WEISBERGER SUM RULE

Before evaluating the σ-model axial charge, let us try to understand the agreement between experiment and the Adler-Weisberger sum rule. Weinberg's derivation[14] of this sum rule has the following steps:

▷ The isovector πN scattering length is determined using soft pion techniques (based on PCAC, of course) to be

$$a_1 = a_{3/2} - a_{1/2} \approx \frac{3m_\pi}{8\pi f_\pi^2(1 + m_\pi/M)} \tag{9}$$

▷ Eq. 9 can be expressed as a dispersion integral over πN cross sections[15] (which we need not exhibit here).

▷ The Goldberger-Treiman relation[16],

$$f_\pi = \frac{g_A M}{g} \tag{10}$$

substituted in Eq. 9 then enables us to express g_A^{-2} either as

$$g_A^{-2} = \frac{8\pi a_1}{3} \frac{M^2}{g^2 m_\pi} \tag{11}$$

or as a dispersion integral (Adler-Weisberger sum rule).

If the isovector scattering length a_1 appearing in Eq. 11 were evaluated in the 1-pole approximation, g_A would reduce to unity. That is, the Adler-Weisberger relation (or Goldberger-Miyazawa-Oehme[15] relation) can be represented diagrammatically as in Fig. 1 below. The axial charge renormalization then results explicitly from the rescattering diagrams, Fig. 1c, *etc*.

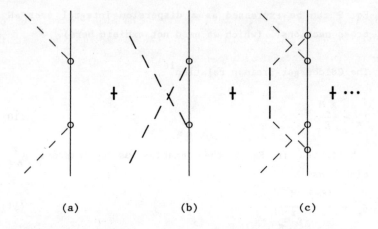

$$+ \qquad \qquad + \qquad \qquad + \qquad \qquad + \cdots$$

$$(a) \qquad \qquad (b) \qquad \qquad (c)$$

Fig. 1 *Pion-nucleon scattering amplitudes*

The above discussion makes clear that the crucial part of the Adler-Weisberger sum rule is the Goldberger-Treiman relation, Eq. 10. As is well-known[17], the latter follows immediately from PCAC together with the assumption that the πNN vertex is slowly varying in momentum-transfer. It therefore seems plausible that the simplest approach to g_A lies within the framework of vertex renormalization.

5. VERTEX RENORMALIZATION IN THE LINEAR σ-MODEL

The Lagrangian density of the linear σ-model has the form[9]

$$\mathcal{L} = \bar{N}\left[i\gamma^\mu\partial_\mu - ig\vec{\tau}\cdot\vec{\pi}\gamma^5 - g\sigma\right]N + \frac{1}{2}\partial_\mu\sigma\partial^\mu\sigma + \frac{1}{2}\partial_\mu\vec{\pi}\partial^\mu\vec{\pi}$$

$$- \frac{\lambda}{4}\left((\sigma^2+\vec{\pi}^2) - f_\pi^2\right)^2 - c\sigma \tag{12}$$

The explicit symmetry breaking term $c\sigma$ gives the pion a mass, and can be identified as $m_\pi^2 f_\pi \sigma$. The axial current can be identified *via* the local infinitesimal transformation

$$N \rightarrow \left[1 + i\frac{1}{2}\gamma^5\vec{\tau}\cdot\vec{v}\right]N \tag{13a}$$

$$\vec{\pi} \rightarrow \vec{\pi} - \vec{v}\sigma \tag{13b}$$

$$\sigma \rightarrow \sigma + \vec{v}\cdot\vec{\pi} \tag{13c}$$

to be

$$\vec{A}^\mu = \bar{N}\frac{1}{2}\vec{\tau}\gamma^\mu\gamma^5 N + \sigma\partial^\mu\vec{\pi} - \vec{\pi}\partial^\mu\sigma \ . \tag{14}$$

Equation 14 is partially conserved:

$$\partial_\mu \vec{A}^\mu = m_\pi^2 f_\pi \vec{\pi}(\vec{x}) \ , \tag{15}$$

which might lead us to suppose that in the limit of vanishing pion mass the axial current will not be renormalized by strong interactions (because it is manifestly conserved): naively,

$$\lim_{m_\pi \rightarrow 0} g_A = 1 \ . \tag{16}$$

However, Eq. 16 is correct only in the Wigner mode. If the vacuum is infinitely degenerate (Goldstone mode) then the axial charge is renormalized even in the massless pion limit[18].

At zero momentum transfer, the only $O(g^2)$ loop diagrams that renormalize g_A are those shown in Fig. 2 below. It is straightforward to see that the logarithmic (ultraviolet) divergence in Fig. 2b cancels that in Fig. 2c, so the sum of Fig. 2b and Fig. 2c is finite and needs no further renormalization. The contribution from the diagram shown in Fig. 3 below (arising from the explicit mesonic term in Eq. 14) is easily seen both to be finite, and to vanish at $q^2 = 0$.

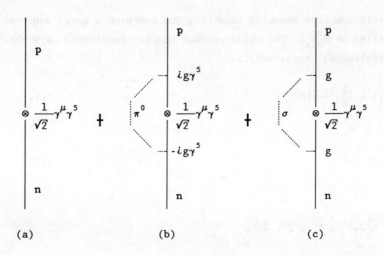

Fig. 2 *σ-model loop diagrams contributing to g_A to $\mathcal{O}(g^2)$*

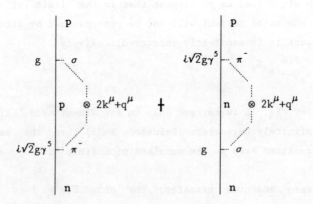

Fig. 3 *Explicit meson contribution to g_A*

In order to evaluate the diagrams Fig. 2b,c we impose a cutoff at the meson-nucleon vertices and take $q^2 = 0$. The cutoff is taken to be a function of a scalar that reduces to the square of the meson 3-momentum in the rest frame of the (emitting or absorbing) on-shell nucleon, namely

$$K^2 = -\left[k - p\frac{p \cdot k}{M^2}\right]^2 .$$ (17)

In fact, it is convenient to take

$$v(K^2) = \exp(-\lambda K^2/M^2)$$ (18)

since we may then apply the exponentiation technique[19] for combining propagators, without introducing extra parameters to account for the cutoffs. The result for the diagrams Fig. 2b,c is

$$\Delta g_A^{(2)} = \frac{g^2}{16\pi^2} \int_0^1 dx\, x \int_0^\infty dw \left[\frac{w}{w+2\lambda}\right]^{3/2} e^{-wx^2} .$$

(19)

$$\left\{\left[(2-x)^2 - \frac{w+\lambda}{w(w+2\lambda)}\right]\exp\left[-\frac{m_\sigma^2}{M^2}(1-x)w\right] - \left[x^2 - \frac{w+\lambda}{w(w+2\lambda)}\right]\exp\left[-\frac{m_\pi^2}{M^2}(1-x)w\right]\right\}$$

giving the values in Table I below. The cutoff parameter range that yields anomalous magnetic moments in reasonable agreement with experiment[20] (indicated by arrows in Table I) predicts an axial renormalization much larger than the experimental value ($\Delta g_A \approx 0.67$). (It is worth noting that the cutoff range is consistent with the "little bag"[21] but not with the MIT bag or its chiral variants[22].)

Table I

Second-order σ-model contributions to g_A
as a function of cutoff range

λ	$\Delta g_A^{(2)}(\lambda)$	
0.00	1.46	
0.01	1.35	
0.05	1.17	
0.10	1.03	
$\rightarrow$ 0.25	0.78	$\leftarrow$
$\rightarrow$ 0.50	0.57	$\leftarrow$
1.00	0.38	

The values of λ corresponding to anomalous magnetic moments
in agreement with experiment are indicated by arrows $\rightarrow$

The free parameters are $g^2/4\pi = 14$; $m_\sigma \approx 700$ MeV/c^2

We may conclude that the axial renormalization predicted by the linear σ-model with cutoff, as well as that given by chiral bag models (which share the same underlying physics), disagrees with that found empirically. In other words, PCAC is not enough — there is some missing physics.

6. AXIAL CHARGE IN THE $\sigma+\omega$ MODEL

After some pondering I concluded that the missing ingredient had to be the same vector mesons invoked[23] to explain the anomalous magnetic moments of nucleons. Since the ρ meson is relatively weakly coupled to the nucleon, the only important candidate seemed to be the ω meson, which contributes through vertex corrections such as that shown in Fig. 4 below. As Preparata and Weisberger have shown[24], the axial renormalization is finite in a renormalizable theory that respects PCAC, once charge- and wave function renormalizations have been carried out. This is the case in the present calculation: the divergence in the wave function renormalization diagram, Fig. 4b cancels that in the vertex term, Fig. 4a.

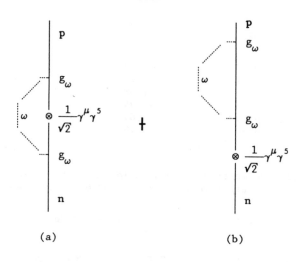

(a)   (b)

Fig. 4 *Vector meson contributions to* g_A:

   (a) *logarithmically divergent vertex correction*
   (b) *logarithmically divergent wave function*
      renormalization

The result, using the same parameterization and vertex cutoff functions as in Eq. 19 above, is

$$\Delta g_{A,\omega}^{(2)} = \frac{-g_\omega^2}{8\pi^2} \int_0^1 dx \int_0^\infty dw \left[\frac{w}{w+2\lambda}\right]^{3/2} \exp\left[-wx^2 - \frac{m_\omega^2}{M^2}(1-x)w\right] \cdot$$

(20)

$$\left\{x\left[1 + (1-x)^2 - \frac{w+\lambda}{w(w+2\lambda)}\right] + \frac{1-x}{w}\right\}$$

We display the corrections given by Eq. 20 in Table II below:

Table II

Second-order ω-loop contributions to g_A
as a function of cutoff range

λ	$\Delta g^{(2)}_{A,\omega}(\lambda)$	
0.00	-1.32	
0.01	-1.21	
0.05	-1.08	
0.10	-0.94	
→ 0.25	-0.69	←
→ 0.50	-0.49	←
1.00	-0.32	

The values of λ corresponding to anomalous magnetic moments
in agreement with experiment are indicated by arrows →

The free parameters are $g^2_\omega/4\pi = 8$; $m_\omega = 783$ MeV/c^2

We see that the correction from incorporating the vector mesons in the theory is substantial, and in the right direction to resolve the excessively large g_A predicted by the bag- or linear σ-model. Depending on what we assume for the bare axial charge (0.84-1.18 in the earlier estimates) and on what value we take for the σ mass, we can obtain the observed axial charge of the nucleon (≈ 1.28) with a range of cutoff parameters λ and ωNN coupling constants g_ω consistent with those found in NN scattering and other low- to medium energy processes[25].

7. NUCLEAR RENORMALIZATION OF AXIAL CHARGE

One of the most obvious questions arising in calculations of axial charge based on structural models of the hadrons is whether the nuclear environment further modifies the axial charge. As shown previously in connection with magnetic moment renormalization effects[20], the nuclear environment manifests itself in the following ways:

▷ Pauli exclusion of intermediate-state nucleons from occupied levels (exchange currents[26]);

▷ Modifications of the meson and baryon propagators:

   O by meson-baryon scattering;

   O by mass changes induced by partial restoration of chiral symmetry.

The Pauli modifications can be estimated (at $q^2 = 0$) by replacing the nucleon propagators in Fig. 2b,c and 4a,b by[20]

$$S_F(p) \rightarrow \frac{1}{\gamma \cdot p - M + i\eta} + 2\pi i \delta \left[p^0 - \sqrt{\vec{p}^2 + M^2} \right] \theta(k_F - |\vec{p}|) \frac{\gamma \cdot p + M}{2p^0} \qquad (21)$$

The pion diagram, Fig. 2b, is modified by

$$\Delta g_A^\pi \approx \frac{-g^2}{24\pi^2} \int_0^{k_F} dk k^4 \left[M^2 + k^2 \right]^{-3/2} \left[m_\pi^2 + k^2 \right]^{-1} \approx -0.003 \qquad (22)$$

where a Fermi momentum $k_F = 250$ MeV/c was assumed.

The ω and σ loop contributions are approximately

$$\Delta g_A^{\sigma,\omega} \approx \frac{\mp g^2}{2\pi^2} M^2 \int_0^{k_F} dk k^2 \left[M^2 + k^2 \right]^{-3/2} \left[m^2 + k^2 \right]^{-1} \qquad (23)$$

where the - sign apppends to the σ diagram, and the appropriate meson mass m and coupling constant g is intended. The sum of the results is about -0.05 or less, depending on the ωNN coupling constant.

In other words, the Pauli quenching of g_A is fairly small. The effect of mass changes can be estimated by the sensitivity to the cutoff parameter. For the range of values that agrees with magnetic moments of free baryons, g_A is insensitive to the cutoff, hence we expect that reasonable mass shifts will also have little effect.

Finally, what about rescattering corrections? As we found in the study of magnetic moments[20], the only serious rescattering effect arises from the p-wave π-nucleon amplitude, which modifies the pion propagator to

$$\Delta_\pi(k) = \left[(k^0)^2 - \vec{k}^2 [1-c\rho v^2(\vec{k})] - m_\pi^2 + i\eta \right]^{-1} \qquad (24)$$

where $-c\vec{k}^2 v^2(\vec{k})$ is the forward pion scattering amplitude for a single nucleon and ρ is the local nuclear density. At nuclear matter density, $\mathcal{R}e\left[c\rho v^2(0) \right] \approx 1$, hence there is a limited region of $\vec{k}^2$ where the pion propagator is much larger than its isolated-pion value. This effect can be studied in the pion term of Eq. 19 by increasing m_π, increasing λ (to restrict the pion 3-momentum to small values) and multiplying by a factor $\approx (1-c\rho)^{-1} - 1$. The net effect is, again, small, as we found previously for isovector anomalous magnetic moments[20].

8. <u>CONCLUSIONS</u>

The results of this investigation can be summarized as follows:

▷ Bag models consistent with the n-p mass difference predict g_A in the range 0.8-1.6, consistent with the "bare-nucleon" value $g_A = 1$ as well as the nonrelativistic limit $g_A = \frac{5}{3}$.

▷ The σ-model value of g_A is too large, even with realistic cutoffs at the meson-baryon vertex;

▷ The corrections found by incorporating the ω meson are negative and sufficiently large to represent the physical ingredient needed to correct the σ-model;

▷ With realistic cutoff parameters (that agree with the anomalous magnetic moments) the $\sigma+\omega$ model axial charge is insensitive to the radius of the hadronic charge distribution;

▷ The effects of exchange corrections (Pauli corrections to intermediate nucleon propagators) and πN rescattering in the nuclear medium are small;

▷ Hence, g_A should not be significantly renormalized in nuclei.

9. ACKNOWLEDGEMENTS

I am grateful to the Summer Institute in Theoretical Physics (Queens University, Kingston, Ontario, Canada) for hospitality during the course of this work; and to Professor H. Leutwyler for several useful discussions.

REFERENCES

[1] G.E. Brown and M. Rho, Phys. Lett. **82B** (1979) 177;
G.E. Brown, M. Rho, and V. Vento, Phys. Lett. **84B** (1979) 383.

[2] S.L. Adler, Phys. Rev. Lett. **14** (1965) 1051.
W. Weisberger, Phys. Rev. Lett. **14** (1965) 1047.

[3] S. Weinberg, in *1970 Brandeis University Summer Institute in Theoretical Physics, v. 1* (MIT Press, Cambridge, 1970).

[4] R. Tegen, Phys. Lett. **172B** (1986) 153;
R. Tegen, P. Zimak and R.D. Viollier, J. Phys. G: Nucl. Phys. 12 (1986) L243.

[5] See, *e.g.*, S. Weinberg, in *A Festschrift for I. Rabi*, ed. L. Motz (New York Academy of Sciences, New York, 1977) p. 185.

[6] See, *e.g.*, P. Simic, Phys. Rev. **D34** (1986) 1903;
C.M. Fraser, Z. Phys. **C28** (1985) 101.

[7] G. Adkins, C.R. Nappi and E. Whitten, Nucl. Phys. **B228** (1983) 552;
G. Adkins and C.R. Nappi, Nucl. Phys. **B233** (1983) 552.

[8] R. Friedberg and T.D. Lee, Phys. Rev. **D15** (1977) 1694; *ibid.* D16 (1977) 1096.
M. Araki, Phys. Rev. **C35** (1987) 1954.

[9] M. Gell-Mann and M. Levy, Nuovo Cimento **16** (1960) 53.

[10] S. Weinberg, *op. cit.*, Ref. 3, p. 382.

[11] S. Adler, Phys. Rev. 137 (1965) B1022; **139** (1965) B1638.

[12] This means that — as the ω is massive — its elementary coupling must be to (conserved) baryon number. See, *e.g.*, W. Zimmermann in *1970 Brandeis University Summer Institute in Theoretical Physics*, v. 1 (MIT Press, Cambridge, 1970).

[13] M. Araki, *op. cit.*, Ref. 8.

[14] See, *e.g.*, S. Weinberg, *op. cit.*, Ref. 3.

[15] M.L. Goldberger, H. Miyazawa and R. Oehme, Phys. Rev. **99** (1955) 986.

[16] M.L. Goldberger and S.B. Treiman, Phys. Rev. **110** (1958) 1178.

[17] M. Gell-Mann and M. Lévy, *op. cit.*, Ref. 9.

[18] This is obvious from the Adler-Weisberger relation. I am indebted to Professor W.I. Weisberger for identifying the spontaneous symmetry breaking as the reason why Eq. 16 is false.

[19] J.D. Bjorken and S. Drell, *Relativistic Quantum Mechanics* (McGraw Hill Book Company, NY, 1964), Eq. 8.12.

[20] J.V. Noble, Nucl. Phys. **A368** (1981) 477.

[21] G.E. Brown and M. Rho, *op. cit.*, Ref. 1;
J.V. Noble, in *Windsurfing the Fermi Sea*, v. II (Elsevier Science Publishers, B.V., 1987), p. 83.

[22] R. Jaffe, *Proc. 1979 Erice Summer School "Ettore Majorana"*, ed. A. Zichichi (Plenum, New York, 1981).

[23] Y. Nambu, Phys. Rev. **106** (1957) 1266; W.R. Frazer and J.R. Fulco, Phys. Rev. Lett. **2** (1959) 365.

[24] G. Preparata and W.I. Weisberger, Phys. Rev. **175** (1968) 1965.

[25] M.M. Nagels, *et al.*, Nucl. Phys. **B109** (1976) 1.

[26] H. Miyazawa, Prog. Theor. Phys. **5** (1951) 801.

STUDY OF THE $B = 1$ SOLUTION OF THE NAMBU–JONA-LASINIO MODEL

K.Goeke, Th.Meißner, E.Ruiz Arriola, F.Grümmer, H.Mavromatis

Institut für Kernphysik, Kernforschungsanlage Jülich GmbH.
D-5170 Jülich, West Germany

ABSTRACT

The Nambu–Jona-Lasinio model (NJL) in the chiral invariant SU(2)–sector with scalar couplings is solved numerically in the Hartree approximation (zero boson loop) for baryon number $B = 1$. To this end first the polarized vacuum solution $(B = 0)$ is constructed using appropriately parametrized non–dynamic meson fields on the chiral circle. The cut–off Λ is fixed to reproduce the pion decay constant. With this choice a full treatment of the polarized vacuum is shown in second order gradient expansion to be equivalent to considering kinetic energies of the mesons. Solutions of the NJL model with baryon number $B = 1$ are obtained by adding $N_c = 3$ valence quarks to the full polarized vacuum and subjecting them to the same meson fields. If one adds the valence quarks to the kinetic energy of the mesons the usual chiral soliton model with valence quarks (CSM) is obtained. For both, NJL and CSM, the equilibrium radii of the $B = 1$ solution are evaluated and shown to be rather close to each other. The present approach shows no vacuum instabilities. The resulting radii are different from those of the renormalized one quark loop model.

There are various indications[1-7] that the Nambu–Jona-Lasinio model[8] plays an intermediate role between low energy QCD and chiral effective meson and quark–meson Lagrangians. It is the purpose of the present paper to study this model in the Hartree approximation (zero boson–loop and one fermion–loop). Thereby we will focus our attention on the question how to construct a solution with baryon number one (B=1) and how far a chiral soliton model with valence quarks (Gell-Mann–Levi Lagrangian[9]), as it has frequently and successfully been applied in the last years[10-16], can be related to the Nambu–Jona-Lasinio model via a gradient[17-19] or heat kernel[20] expansion. This relationship will be studied conceptually and numerically and it will turn out that the chiral soliton model seems to be a very well defined and good approximation to the Nambu–Jona-Lasionio model, if the corresponding cut–off is properly chosen. We, furthermore, will contrast the present Nambu–Jona-Lasionio procedure to the renormalized one–quark loop approach to the chiral soliton model of Ripka and Kahana[21], a theory which is known to show vacuum instabilities[22,23].

The Nambu–Jona-Lasionio Lagrangian, as it is used in this paper, reads

$$\mathcal{L}_{NJL} = \overline{\psi}i\gamma^\mu\partial_\mu\psi + \frac{G}{2}[(\overline{\psi}\psi)^2 + (\overline{\psi}i\gamma_5\vec{\tau}\psi)^2] \tag{1}$$

It is not renormalizeable and must be used with a finite cutoff Λ.

The generating functional

$$\mathcal{Z} = \int D\overline{\psi}D\psi \exp\{i\int d^4x \mathcal{L}_{NJL}(x)\}$$

can equivalently be written[20,24,25] as

$$\mathcal{Z}' = \int D\overline{\psi}D\psi D\sigma D\vec{\pi} \exp\{i\int d^4x \mathcal{L}'_{NJL}\}$$

with

$$\mathcal{L}'_{NJL} = \overline{\psi}i\gamma^\mu\partial_\mu\psi - g\overline{\psi}(\sigma + i\gamma_5\vec{\pi}\cdot\vec{\tau})\psi - \frac{1}{2}\mu^2(\sigma^2 + \vec{\pi}^2) \tag{2}$$

and $G = \frac{g^2}{\mu^2}$. Similarly to Dyakonov et. al.[26] the boson fields are considered to be classical. This corresponds to a zero-boson-loop and one-quark-loop treatment. They are not treated self-consistently but are assumed to be on the chiral circle and to be parametrized by means of an appropriate chiral profile $\theta(r)$ as $\sigma = f_\pi \cos\theta(r)$ and $\vec{\pi} = f_\pi \hat{r} \sin\theta(r)$. Due to this choice the third term on the right-hand side of eq. (2) can be absorbed into the energy of the vacuum. As profiles we consider the forms $\theta(r) = \pi \exp(-r/R)$ and $\theta(r) = \pi(1 - r/R)$ with R as a variational parameter. Following the techniques described by Ripka and Kahana[23] the Dirac equation is solved for all positive and negative energy quark levels by putting the system into a finite box.

Using the proper time regularization[20,27] the total energy of the vacuum (baryon number $B = 0$) can be derived to

$$E_0(R) = N_c \left\{ \sum_\lambda R_1(\epsilon_\lambda, \Lambda) - \sum_k R_1(\epsilon_k, \Lambda) \right\} \qquad (3)$$

Here $N_c = 3$ is the number of colours, the ϵ_λ are eigenvalues of $h = -i\vec{\alpha} \cdot \vec{\nabla} + g(\sigma + i\vec{\pi} \cdot \vec{\tau}\gamma_5)$ and $\epsilon_k = \pm\sqrt{(k^2 + g^2 f_\pi^2)}$. The sum has to be extended over positive and negative eigenvalues and $R_1(\epsilon, \Lambda)$ is the regularization function

$$R_1(\epsilon, \Lambda) = (\frac{1}{4\sqrt{\pi}})^{1/2} \int\limits_{1/\Lambda^2}^{\infty} du \, u^{-3/2} \exp(-u\epsilon^2) \qquad (4)$$

Up to now the cut-off parameter Λ has not been fixed. For a consistent treatment the Λ must be chosen such that the pion decay constant f_π is reproduced. This corresponds to demanding for a given quark-meson coupling constant g

$$\langle 0|A_\mu^a|\pi^b(p)\rangle = ip_\mu e^{-ipx} f_\pi(m_\pi = 0)\delta_{ab}$$

,where $\vec{A}_\mu$ is the axial vector current. In the 1-quark-loop approximation this amounts to evaluate the Feynman diagram:

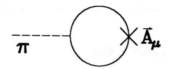

Using the proper time regularization of f_π this yields with $\alpha = (gf_\pi/\Lambda)^2$

$$f_\pi^2 = f_\pi^2 R_2(gf_\pi, \Lambda) = \frac{g^2 f_\pi^2 N_c}{4\pi^2} \int\limits_\alpha^\infty \frac{du}{u} e^{-u} \qquad (5)$$

This simple feature provides a direct link to the chiral soliton models. They use in contrast to Nambu–Jona-Lasionio explicitly the kinetic energy of the mesons. Indeed, if one performs in $\mathcal{Z}'$ a functional integration over the quarks one ends up with the effective Lagrangian

$$\mathcal{L}'_{NJL} = -i N_c Tr\langle x| \ln \left[i\gamma^\mu \partial_\mu - g(\sigma + i\gamma_5 \vec{\pi} \cdot \vec{\tau}) \right] |x\rangle \qquad (6)$$

where Tr indicates a trace over spinor and isospin degrees of freedom.

If we perform, furthermore, a gradient expansion[17-20] up to second order we obtain for the energy of the polarized vacuum

$$E_0^G(R) = \frac{1}{2} R_2(gf_\pi, \Lambda) \int d^3x \left[(\vec{\nabla}\sigma)^2 + (\vec{\nabla}\pi)^2 \right] \qquad (7)$$

The baryon number of the polarized vacuum is zero as long as the valence orbit has a positive eigenvalue.

It is important to note that with the above choice of Λ, obtained by demanding eq. (5) to be fulfilled, one gets $R_2(gf_\pi, \Lambda) = 1$ and thus E_0^G is identical to the kinetic energy of the meson fields. Hence this kinetic energy term represents the polarization of the vacuum caused by the non-dynamic meson fields. Thus for actual calculations one can use either of them as long as the second order gradient approximation is a good one,

48

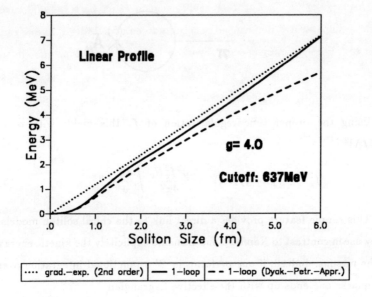

Fig. 1: *Using the linear profile the total vacuum energy $(E_0(R)$, eq. (3), solid line), its second order gradient expansion $(E_0^G(R)$, eq. (7), kinetic meson energy, dashed line) and the Dyakonov-Petrov approximation to $E_0(R)$ (dotted) are presented. The calculations are done with the parameters $g = 4$ and $\Lambda = 637 MeV$.*

which will be investigated below.

That indeed the kinetic energy of mesons is generated by the polarization of the vacuum is seen explicitly in Fig. 1 and provides a good check of the numerical accuracy. For profiles $\theta(r)$ being functions of r/R the $E_0^G(R)$ of eq. (7) is linear in R, i.e. $E_0^G(R) = aR$ and the value of a depends on the profile chosen and the values of g and Λ. If one evaluates the total energy of the vacuum with the same profile and the same parameters one obtains

$$E_0(R) = aR + \frac{b}{R} + \frac{c}{R^3} + \cdots$$

Thus for large R the kinetic meson energy and the exact $E_0(R)$ should have the same slope, if Λ is determined according to eq. (5). Indeed, as one can see at Fig. 1, the numerical accuracy of our calculations is high enough to ensure this. For comparison we present also the total vacuum energy in an approximation suggested by Dyakonov et.al.[26], whose basic conceptual arguments are rather similar to ours. Their approximation consists in expanding $R_1(\epsilon_\lambda, \Lambda)$ of eq. (3) linearly around ϵ_λ and assuming tr $h = 0$ even for finite Λ. For $R = 6fm$ their approximation to $E_0(R)$ yields only 80% of the kinetic energy and the deviations increase with increasing R. It is, however, fair to say that the equilibrium radii of the $B = 1$ solutions, to be discussed below, are only little affected.

We are now in the position to construct solutions with $B = 1$. They are obtained by adding $N_c = 3$ quarks to the system and subjecting them to the same boson fields. If we denote the corresponding quark eigenvalue with $\epsilon_{val}(R)$ the total energy of the $B = 1$ solution is

$$E_1(R) = N_c \epsilon_{val}(R) + E_0(R) \tag{8}$$

This formula is correct for the range of R where $0 \leq \epsilon_{val}(R) < gf_\pi$ which is relevant for the following considerations.

The curves for $N_c \epsilon_{val}(R)$ and $E_0(R)$ are given for the exponential profile in Fig. 2. The $E_1(R)$ exhibits a local minimum indicating a stable solution of the Nambu–Jona-Lasionio model. In the second order gradient approximation one considers the energy

$$E_1^G(R) = N_c \epsilon_{val}(R) + E_0^G(R) \tag{9}$$

Since with the chosen cut-off we have $E_0^G = (\vec{\nabla}\sigma)^2/2 + (\vec{\nabla}\pi)^2/2$, eq. (9) is exactly equal to the expression minimized in all those papers which study the linear chiral soliton model in the valence quark approximation[9-16] if one works on the chiral circle and in the soft pion limit ($m_\pi = 0$). Thus

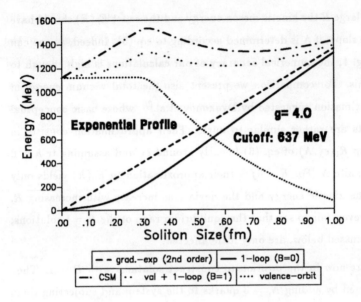

Fig. 2 : Using the exponential profile the total vacuum energy ($E_0(R)$, solid line), the kinetic meson energy ($E_0^G(R)$, short dashed line), the valence quark eigenvalue $N_c\epsilon_{val}$ (dotted line) and the total energies for the $B = 1$ solution $E_1 = N_c\epsilon_{val} + E_0(R)$ (long dashed) and $E_1^G = N_c\epsilon_{val} + E_0^G(R)$ (dashed double-dotted) are presented. The calculations are done with $g = 4$ and $\Lambda = 637 MeV$.

these models appear as clearly defined approximations to the Nambu–Jona-Lasionio model in the zero-boson- and one-quark-loop approximation.

It is interesting to investigate how good this valence quark approach really is. For this we have calculated with the exponential profile the position of the minima of $E_1(R)$ and of $E_1^G(R)$ and presented them in Tab. 1. Here two situations are considered, one for fixed g and the other, where g is adjusted such that all approaches yield the same equilibrium energy. The latter procedure corresponds to treating g as a free parameter to be adjusted to observables. From Tab. 1 one concludes that as far as the sizes are concerned the valence-quark approach to the chiral soliton model is a good approximation to the Nambu–Jona-Lasinio model in the $B = 1$ sector. Furthermore, if the quark–meson coupling strength g is readjusted, then the gradient (or heat kernel) expansion to second order is sufficient and higher orders do not contribute much. The same conclusions can be drawn from the results with the linear profile.

	g	$\Lambda[MeV]$	$E_{min}[GeV]$	$R_{min}[fm]$
$E_1 = $ Sea +Val.	4	637	1.130	0.6
$E_1^G = $ Kin.Mes.+Val.	4	–	1.330	0.7
Ripka and Kahana	4	∞	1.070	0.4
$E_1 = $ Sea +Val.	4.8	638	1.072	0.7
$E_1^G = $ Kin.Mes.+Val.	5.5	–	1.079	0.8
Ripka and Kahana	4	∞	1.070	0.4

Tab. 1: *Using the exponential profile the minimum of the full solution (E_1, eq. (8)) and the second order gradient expansion (E_1^G, eq. (9)) are identified and compared with the renormalized 1-quark-loop approach of Ripka and Kahana (Ref. 23). In the lower part g is adjusted to yield roughly identical energies.*

The curves of $E_1(R)$ and $E_1^G(R)$ in Fig. 1 deserve some further explanation. They are only meaningful for a range of soliton sizes R in which $0 \leq \epsilon_{val}(R) < gf_\pi$. In this range we have baryon number $B = 1$ and eq. (8) holds. If $\epsilon(R)$ becomes negative at around $1.2\,fm$ the valence quark contribution to the energy is already contained in $E_0(R)$ and eq. (8) is no longer valid. As a matter of fact the minima occur at $R \approx 0.6 - 0.7\,fm$ and our present procedure is correct. Another problem occurs at small R, because in principle the system may tunnel from the local minimum to $R = 0$, where the energy is lower if one takes Fig. 1 literally. We do not know if this is really a problem, because a fully selfconsistent calculation without a prescribed chiral profile may perhaps yield noticeably lower energies for the local minima. Furthermore, it is not clear at all how a valence quark picture can be formulated for the case where the valence quark level has entered the continuum. Thus probably the curves of the $B = 1$ solution in Fig. 1 are meaningless for $R < 0.3 fm$.

It is interesting to compare the present concepts and numbers with the model of Ripka and Kahana[23]. These authors solve the Lagrangian of eq. (2) on the chiral circle completed by an explicit kinetic energy term of the mesons yielding in fact the old Lagrangian of Gell-Mann and Levy. They treat this Lagrangian after renormalization in the one-quark loop approximation. In our framework this consists in adding counterterms to $E_1(R)$ in eq. (8) and considering the limit $\Lambda \to \infty$. A comparison of their equilibrium radius[23] with the present ones both obtained with an exponential profile is performed in Tab. 1. Although it is conceptually clear that the model of Ripka and Kahana differs from the present approach[26] it is, nevertheless, interesting to see that this is also reflected in actual numbers. It turns out that the stabilization radii are much smaller and hence the chiral soliton model with valence quarks is not an approximation to Ripka and Kahana's model. This fact has some interesting consequences. As discussed by various authors[22,23] the Ripka–Kahana model shows an instability of the translationally invariant vacuum at $R \simeq 0.2 fm$. This feature does not occur in the present theory because by construction the cut-off Λ is finite.

To summarize our points: We have shown that the chiral valence quark soliton model can be considered as an approximation to the Hartree-solution (zero-boson and one-fermion loop) of the Nambu–Jona-Lasinio model. This requires a heat kernel or gradient expansion with a proper time regularization and a cut-off which is chosen to reproduce the experimental value of the pion decay constant. Then the equilibrium radii of the chiral soliton model and the Nambu–Jona-Lasionio model agree within 15%. Vacuum instabilities are not encountered.

Although with the present profiles the conclusions seem to be rather clear, more detailed investigations are necessary which avoid the chiral circle, allow for a finite pion mass and, in particular, are selfconsistent. In addition, regularization effects of the baryon number should be studied. If the present conclusions will be approved and if indeed, as suggested by many authors, the Nambu–Jona-Lasinio model turns once out to be a reliable approximation to low energy QCD, then one would have a direkt link from QCD to the chiral valence quark soliton model.

The paper has been supported in parts by the NATO grant RG85/0217. Useful discussions with G.E.Brown, M.Harvey and J.N.Urbano are acknowledged. Th.Meißner likes to thank the "Studienstiftung des deutschen Volkes" for financial support.

REFERENCES

1. R.D.Ball, *Desparately seeking mesons*, Workshop on Skyrmions and Anomalons, Kracov 1987, Publ. World Scientific

2. D.Dyakonov, Yu.V.Petrov, *Nucl. Phys.* **B 272**, 457 (1986)

3. D.McKay, H.Munczek, *Phys. Rev.* **D 32**, 266 (1985)

4. P.Simic, *Phys. Rev.* **D 34**, 1903 (1986)

5. R.Cahill, C.Roberts, *Phys. Rev.* **D 32**, 2419 (1985)

6. A.Adrianov, *Phys. Lett.* **157 B**, 425 (1985)

7. H.Reinhardt, *QCD condensates and effective quark Lagrangians*, NBI–preprint, 1986

8. Y.Nambu, G.Jona-Lasinio, *Phys. Rev.* **122**, 354 (1961)

9. M.Gell-Mann, M.Levi, *Nuov. Cim.* **16**, 705 (1960)

10. S.Kahana, G.Ripka, V.Soni, *Nucl. Phys.* **A 415**, 351 (1984)

11. M.C.Birse, M.K.Banerjee, *Phys. Lett.* **136 B**, 284 (1984)

12. M.C.Birse, *Phys. Rev.* **D 33**, 1934 (1986)

13. M.Fiolhais, A.Nippe, K.Goeke, F.Grümmer, J.N.Urbano, *Phys. Lett.* **194 B**, 187 (1987)

14. K.Goeke, M.Harvey, F.Grümmer, J.N.Urbano, *Phys. Rev.* **D 37**, 754 (1988)

15. M.Fiolhais, K.Goeke, F.Grümmer, J.N.Urbano, *Nucl. Phys.* **A 481**, 727 (1988)

16. P.Alberto, E.Ruiz Arriola, M.Fiolhais, F.Grümmer, J.N.Urbano, K.Goeke, *Phys. Lett.* **208 B**, 75 (1988)

17. J.A.Zuk, *Z. Phys.* **C 29**, 303 (1985)

18. L.H.Chan, *Phys. Rev. Lett.* **55**, 21 (1085)

19. J.J.R.Aitchison, C.M.Frazer, *Phys. Rev.* **D 31**, 2608 (1985)

20. D.Ebert, H.Reinhardt, *Nucl. Phys.* **B 271**, 188 (1986)

21. S.Kahana, G.Ripka, *Nucl. Phys.* **A 429**, 462 (1984)

22. V.Soni, *Phys. Lett.* **195 B**, 569 (1987)

23. G.Ripka, S.Kahana, *Phys. Rev.* **D 36**, 1233 (1987)

24. D.Ebert, M.K.Volkov, *Z. Phys.* **C 16**, 205 (1983)

25. T.Eguchi, *Phys. Rev.* **D 14**, 2755 (1976)

26. D.J.Dyakonov, V.Yu.Petrov, P.V.Pobylitsa, *Nucl. Phys.* **B 306**, 809 (1988)

27. J.Schwinger, *Phys. Rev.* **82**, 664 (1951)

Phase Transitions

ON THE CHIRAL PHASE TRANSITION

H. Leutwyler

Institute for Theoretical Physics
University of Bern
Sidlerstrasse 5, CH-3012 Bern, Switzerland

In the chiral limit, the first few terms in the low temperature expansion of QCD can rigorously be calculated, the contributions generated by massive states being suppressed exponentially. Using the dilute gas approximation to account for these contributions, an estimate of the critical temperature is obtained.

1.   THE QUARK CONDENSATE AS AN ORDER PARAMETER

At low temperatures, the behaviour of any system is controlled by its lightest excitations. The lightest strongly interacting particle is the pion. In the following, I wish to show that the low temperature properties of the strong interaction can indeed be predicted on the basis of what we know about the pion. The talk is based on work done in collaboration with Jürg Gasser and with Peter Gerber[1-6].

For definiteness, I discuss the problem in the framework of QCD. What really counts in what follows is that this theory exhibits an approximate, spontaneously broken symmetry[7]; I will extensively exploit the fact that the Hamiltonian of QCD can be split into two parts

$$H = H_0 + H_1 \qquad (1.1)$$

Work supported in part by Schweizerischer Nationalfonds

where H_1 is the mass term of the u and d quarks. The key ingredients in our analysis are the following properties of H_0 and H_1:

(i) H_0 is invariant under the group $SU(2)_R \times SU(2)_L$ of global chiral rotations.

(ii) Chiral symmetry is spontaneously broken, the ground state of H_0 being symmetric only under the subgroup $SU(2)_{R+L}$.

(iii) H_1 can be treated as a small perturbation; it transforms according to the representation $D^{(1/2, 1/2)}$ of the chiral group.

Other features of QCD, such as the fact that in this theory the interaction is mediated by a colour gauge field or that the strange quark is not heavy, either, are not essential here. For simplicity, I ignore the small isospin breaking effects generated by the mass difference and set $m_u = m_d = m$,

$$H_1 = \int d^3x \, m \, \bar{q}q \; ; \qquad \bar{q}q = \bar{u}u + \bar{d}d \qquad (1.2)$$

In QCD, the quark mass is a free parameter which plays a role analogous to an external magnetic field: the quark mass term breaks $SU(2)_R \times SU(2)_L$, like an external magnetic field breaks rotation invariance. The thermodynamic variable conjugate to m is the quark condensate ($\hbar = c = k = 1$)

$$\langle \bar{q}q \rangle = \frac{Tr\{\bar{q}q \, e^{-H/T}\}}{Tr\{e^{-H/T}\}} \qquad (1.3)$$

At zero temperature, the condensate coincides with the vacuum expectation value $\langle 0|\bar{q}q|0 \rangle$. If the ground state of H_0 were invariant under chiral transformations, the vacuum expectation value would vanish in the chiral limit ($m = 0$). In an asymmetric ground state there is no reason for the condensate to vanish - it represents an order parameter associated with spontaneously broken chiral symmetry. In the above analogy, the condensate corresponds to the magnetisation, the chiral limit corresponds to the absence of an external magnetic field and the persistence of the condensate in this limit corresponds to the occurrence of spontaneous magnetisation. As it is the case with the magnetisation, the condensate decreases if the temperature rises. Once the temperature exceeds a critical value, disorder wins, chiral symmetry is restored and the condensate disappears. Presumably, but not necessarily, this happens at the same temperature where the quarks and gluons are liberated, the hadronic plasma being transformed into a quark gluon-plasma.

The spontaneous breakdown of a symmetry gives birth to Goldstone bosons[8] - magnons in the case of spontaneous magnetisation, pions in the case of spontaneously broken chiral symmetry. In the chiral limit, the pions are massless. If the quark mass is turned on, the pions pick up mass. A well known low energy theorem of current algebra[9] states that, for small quark mass, M_π^2 is proportional to the product of the quark mass and of the order parameter

$$M_\pi^2 = -\frac{m}{F_\pi^2} \langle 0|\bar{q}q|0\rangle \{1 + O(m)\} \qquad (1.4)$$

($\langle 0|\bar{q}q|0\rangle$ is negative if $m > 0$; I use the normalization $F_\pi \simeq 93$ MeV for the pion decay constant.) This relation shows that the properties of the lightest excitations occurring in the spectrum of QCD are intimately related to the presence of a symmetry which is broken both spontaneously ($\langle 0|\bar{q}q|0\rangle \neq 0$) and explicitly ($m \neq 0$).

In the thermal equilibrium, the behaviour of the system can be studied by analyzing the properties of the partition function

$$e^{-A/T} = Tr\{e^{-H/T}\} = \sum_n e^{-E_n/T} \qquad (1.5)$$

For the sum over the eigenstates of the Hamiltonian to make sense, the system must be confined to a box of finite volume V. If the volume is large enough, the free energy A is proportional to V, such that the free energy density tends to a finite limit[10,11]

$$z = \lim_{V \to \infty} \frac{A}{V} \qquad (1.6)$$

The condensate is the derivative of z with respect to the quark mass

$$\langle \bar{q} q \rangle = \frac{\partial z}{\partial m}$$

(1.7)

2.   LOW TEMPERATURE EXPANSION

At low temperatures, the pions dominate the properties of the system, because the Boltzmann factor exponentially suppresses the contributions of massive particles to the partition function. In the chiral limit, the pions are massless. The multipion states contribute powers of T and log T

$$z = \sum_{m,n = 0,1,\ldots} C_{mn} (T^2)^m (T^2 \log T)^n + O(e^{-M_1/T})$$

(2.1)

Particles which remain massive in the chiral limit generate contributions of order $\exp(-M_1/T)$, where M_1 is the mass of the lightest such particle. In the limit m_u, $m_d \to 0$, the lightest massive states are the K and η mesons. We could make use of the fact that the properties of these particles can also be understood in terms of a spontaneously broken symmetry. The corresponding symmetry breaking parameter m_s is however 25 times larger than the mean mass of u and d; the masses of K and η are of the order of 500 MeV. As we will see, the cirtical temperature of the chiral phase transition is considerably smaller. These excitations do therefore not play a prominent role in the phase transition. I will discuss their effect on our analysis together with the effects generated by the other massive states ρ, ω, N,... in section 6.

The essential point in the following is that the low energy properties of the pions are to a large extent fixed by chiral symmetry. In particular, chiral symmetry implies that the interaction among low energy pions is proportional to the square of the energy: at low energies, the strong interaction is weak. If the temperature is sufficiently low, the interaction among the pions can be ignored altogether; the pressure is given by the familiar free bose gas formula

$$z = \varepsilon_0 + \frac{3T}{(2\pi)^3} \int d^3p \, \ln\{1 - e^{-E/T}\} + \cdots$$

$$E = \sqrt{M_\pi^2 + \vec{p}^2} \tag{2.2}$$

where the factor of three accounts for π^+, π^-, π^0 and ε_0 is the energy density of the vacuum. With (1.7), we therefore obtain

$$\langle \bar{q}q \rangle = \frac{\partial \varepsilon_0}{\partial m} + \frac{3}{2} g_1 \frac{\partial M_\pi^2}{\partial m} + \cdots$$

$$g_1 = \frac{1}{(2\pi)^3} \int \frac{d^3p}{E} \cdot \frac{1}{e^{E/T} - 1} \tag{2.3}$$

The derivative of the vacuum energy with respect to the quark mass is the vacuum expectation value of $\bar{q}q$

$$\frac{\partial \varepsilon_0}{\partial m} = \langle 0|\bar{q}q|0\rangle \tag{2.4}$$

(compare (1.7)). Using (1.4), the expression for the condensate can be rewritten in the form

$$\langle \bar{q}q \rangle = \langle 0|\bar{q}q|0\rangle \{1 - \frac{3}{2F_\pi^2} g_1 + \cdots\} \tag{2.5}$$

In the chiral limit, the integrals occurring in (2.2) and in (2.3) can be evaluated explicitly. The pressure is given by the familiar T^4-law

$$z = \varepsilon_0 - \frac{\pi^2}{30} T^4 + \cdots \tag{2.6}$$

and the result for the condensate simplifies to

$$\langle \bar{q}q \rangle = \langle 0|\bar{q}q|0\rangle \{1 - \frac{T^2}{8F^2} + \cdots\} \tag{2.7}$$

where F is the value of the pion decay constant in the chiral limit. We have just derived a low temperature theorem. It states that, in the chiral limit, the expansion of $\langle \bar{q}q \rangle$ in powers of T contains a term of order T^2, whose coefficient is determined by the pion decay constant. The formula (2.7) shows that the condensate melts as the temperature rises. The melting is due to the Goldstone bosons whose number density is controlled by the temperature according to familiar free gas formula

$$n_\pi = \frac{3\zeta(3)}{\pi^2} T^3 = .365 \, T^3 \qquad (2.8)$$

If the quark mass is different from zero, the low temperature expansion is not a simple powers series in T and log T. As shown by the free gas formula (2.2) the partition function then involves nontrivial functions of the ratio M_π/T. In particular, for $T \ll M_\pi$, even the contributions generated by the pions are exponentially small. To analyze the behaviour of the system at temperatures of the order of M_π, we treat both T and M_π as small quantities, allowing the ratio M_π/T to have any value (replace T by λT, replace m by $\lambda^2 m$ and expand in powers of λ). In this generalized sense, the low temperature expansion of the free energy density is a power series of the form (2.1) even for nonzero quark mass; the symmetry breaking merely affects the coefficients c_{mn} which now become nontrivial functions of the ratio M_π/T. In the formula (2.5) for the condensate, e.g., the function g_1 is of the form $T^2 h_1$ where h_1 only depends on M_π/T: the first nontrivial term in the low temperature expansion of the condensate is of order T^2 also if $m \neq 0$. (Note that F_π differs from F only through terms of order m; since m counts like two powers of T, this difference becomes significant only if the low temperature expansion is carried to order T^4.)

3.   EFFECTIVE LAGRANGIAN

To carry the low temperature expansion beyond the first nontrivial term, we need to take the interaction among the pions into account. As pointed out above, the strength of this interaction is proportional to the square of the pion energies. At low temperatures, the typical pion energies are small, of order $E \sim T$. At low temperatures, the interaction can therefore be treated as a perturbation.

A very efficient method for the analysis of the perturbations generated by the interaction is the effective Lagrangian technique[12]. I briefly sketch the main features of this method which, in the present context, is also referred to as chiral perturbation theory. For simplicity, I restrict myself to the chiral limit. The quark and gluon fields of the underlying theory are replaced by a pion field which carries the degrees of freedom associated with the lightest physical excitations of the system. The basic properties of this field are determined by the symmetry groups involved in the spontaneous breakdown which is responsible for the occurrence of light excitations in the first place. The field lives on the group SU(2), i.e., it is represented by a unitary 2x2 matrix $U(x)$ with det $U(x) = 1$. The effective action is a nonlocal functional of this field. In the chiral limit, the effective action is invariant under global chiral rotations of the pion field[15]

$$U(x) \rightarrow V_R U(x) V_L^+ \; ; \quad V_R, V_L \in SU(2) \qquad (3.1)$$

If only pions with small four-momenta are involved, the field $U(x)$ is slowly varying and the effective action can then be expanded in a series of local contributions, involving an increasing number of derivatives of the field (derivative expansion). The effective action takes the form of an integral over a local effective Lagrangian which consists of a series of terms[14]

$$L_{eff} = L^{(0)} + L^{(2)} + L^{(4)} + \cdots \qquad (3.2)$$

where $L^{(0)}$ does not contain any derivatives of U, $L^{(2)}$ is linear in $\Box U$ or quadratic in the first derivatives $\sim \partial_\mu U \partial^\mu U$ etc. (Lorentz invariance only permits an even number of derivatives). Actually, since the effective action is invariant under (3.1), the term $L^{(0)}$ is independent of the pion field, i.e. it represents an uninteresting constant describing the energy density of the vacuum in the chiral limit. The most general expressions for $L^{(2)}$ and $L^{(4)}$ consistent with chiral symmetry are

$$L^{(2)} = \frac{F^2}{4} \langle \partial_\mu U^\dagger \partial^\mu U \rangle$$

$$L^{(4)} = \frac{\ell_1}{4} \langle \partial_\mu U^\dagger \partial^\mu U \rangle^2 + \frac{\ell_2}{4} \langle \partial_\mu U^\dagger \partial_\nu U \rangle \langle \partial^\mu U^\dagger \partial^\nu U \rangle \tag{3.3}$$

where $\langle A \rangle$ stands for the trace of the 2x2 matrix A. Chiral symmetry thus fixes the structure of the leading term $L^{(2)}$, up to a single coupling constant F with the dimension of energy. (F is the value of the pion decay constant in the chiral limit, in accord with the notation used above. For a proof of this statement, see, e.g. 14). The structure of $L^{(4)}$ is fixed up to two dimensionless coupling constants ℓ_1 and ℓ_2, etc.

Once the effective Lagrangian is specified, it is straightforward to calculate S-matrix elements. The constraints $U^\dagger U = 1$, det U = 1 are solved by[16]

$$U(x) = e^{i \frac{\vec{\tau} \cdot \vec{\varphi}(x)}{F}} \tag{3.4}$$

where τ_1, τ_2, τ_3 are the Pauli matrices. The first few terms in the expansion of the effective Lagrangian in powers of $\vec{\varphi}$ are

$$L^{(2)} = \frac{1}{2} \partial_\mu \vec{\varphi} \cdot \partial^\mu \vec{\varphi} + \frac{1}{6F^2} (\vec{\varphi} \cdot \partial_\mu \vec{\varphi})(\vec{\varphi} \cdot \partial^\mu \vec{\varphi})$$

$$- \frac{1}{6F^2} (\partial_\mu \vec{\varphi} \cdot \partial^\mu \vec{\varphi}) \vec{\varphi}^2 + O(\varphi^6)$$

$$L^{(4)} = \frac{\ell_1}{F^4} (\partial_\mu \vec{\varphi} \cdot \partial^\mu \vec{\varphi})^2 + \frac{\ell_2}{F^4} (\partial_\mu \vec{\varphi} \cdot \partial_\nu \vec{\varphi})(\partial^\mu \vec{\varphi} \cdot \partial^\nu \vec{\varphi}) + O(\varphi^6) \tag{3.5}$$

The corresponding tree graph contributions to the $\pi\pi$ scattering amplitude are

$$A(s,t)\Big|_{tree} = \frac{s}{F^2} + \frac{2\ell_1 s^2}{F^4} + \frac{\ell_2}{2F^4} \{ s^2 + (t-u)^2 \} \tag{3.6}$$

The first term, which stems from $L^{(2)}$, is proportional to the square of the external four-momenta, while $L^{(4)}$ generates a term of order p^4. The contributions generated by the higher order terms $L^{(6)}$, $L^{(8)}$,... occurring in the derivative expansion are of order p^6 or higher.

The tree graphs are not the full story, of course. Perturbation theory leads to a unitary S-matrix only if graphs involving loops are included[17]. In $\pi\pi$ scattering, e.g., there are one-loop graphs involving two vertices of $L^{(2)}$; the graphs are proportional to $1/F^4$. The loop integral is divergent. Regularizing by analytic continuation in the dimension d, we obtain

$$A(s,t)\Big|_{\substack{\text{one}\\ \text{loop}}} = \frac{1}{6F^4}\left\{3s^2 J(s) + t(t-u)J(t) + u(u-t)J(u)\right\}$$

(3.7)

where J is the elementary integral

$$J(p^2) = (-i)(2\pi)^{-d}\int d^d q\,\{q^2+i\varepsilon\}^{-1}\{(p-q)^2+i\varepsilon\}^{-1}$$

$$= (4\pi)^{-\frac{d}{2}}\cdot\Gamma\left(\tfrac{4-d}{2}\right)\cdot B\left(\tfrac{d-2}{2},\tfrac{d-2}{2}\right)\cdot(-p^2-i\varepsilon)^{\frac{d-4}{2}}$$

(3.8)

The residue of the pole at d = 4 which occurs in the function $J(p^2)$ is independent of p^2. To remove the pole, it therefore suffices to subtract the value of $J(p^2)$ at some reference scale $p^2 = -\mu^2$. Comparing (3.7) with (3.6), one readily checks that the polynomial involving $J(-\mu^2)$ can be absorbed by renormalizing the two coupling constants ℓ_1 and ℓ_2. The sum of the tree graph contribution (3.6) and of the one-loop contribution (3.7) is therefore finite and unambiguous (up to the values of the renormalized coupling constants). Note that the renormalized contribution of the one-loop graphs is not a polynomial in s and t, but involves logarithms

$$J(p^2) - J(-\mu^2) = -\frac{1}{16\pi^2}\ln\left(\frac{-p^2-i\varepsilon}{\mu^2}\right)$$

(3.9)

In dimensional regularization, all graphs are homogeneous functions of the external momenta. If the momenta are scaled, $p \to \lambda p$, each graph picks up a power of λ. To determine this power, it suffices to count the powers of $1/F$ which occur in the vertices of the graph. In this manner, one readily shows that graphs involving two or more loops only

contribute at order p^6 or higher[14]. The leading contribution in the low
energy expansion of the scattering amplitude is the term $s/F^2 = O(p^2)$,
generated by the tree graphs of $L^{(2)}$. At order p^4, both the one-loop
graphs of $L^{(2)}$ and the tree graphs containing one vertex of $L^{(4)}$ contribute. The remaining graphs only matter if the low energy expansion
is carried to order p^6.

 If the quark mass is different from zero, the effective Lagrangian is not invariant under chiral rotations, because the quark
mass term which occurs in the Hamiltonian of QCD explicitly breaks this
symmetry. The asymmetries generated by the mass term can be brought
under control, provided the quark mass is small in comparison to the
scale of the theory. In this case, the full effective Lagrangian can
be approximated by the first few terms in its Taylor series expansion
in powers of m. The derivative expansion can be applied to the coefficients of the Taylor series, term by term. The net result is that
the effective Lagrangian picks up additional contributions involving
powers of the quark mass. The additional terms break chiral symmetry
and therefore equip the pion with a mass. The power counting underlying
the perturbative analysis of the effective pion field theory goes
through without modifications, if the pion mass occurring in the propagators is treated as a small quantity of order p. Since M_π^2 is of
order m, this implies that the quark mass counts as a small quantity of
order p^2. The main modification brought about by a nonvanishing quark
mass is an increase in the number of coupling constants which occur in
the derivative expansion of the effective Lagrangian.

4. CHIRAL PERTURBATION THEORY OF THE PARTITION FUNCTION
 The effective Lagrangian technique is readily extended from
S-matrix elements to the partition function[1,5]. To calculate the contributions to the trace of exp(- H/T) generated by the multipion states,
one first observes that this quantity represents the analytic continuation of the time evolution perator exp(- itH) to the point t = - i/T.
The functional integral representation for the time evolution operator
is given by

$$\langle U'' | e^{-itH} | U' \rangle = \int [dU] \, e^{i \int dx \, \mathcal{L}_{eff}} \tag{4.1}$$

The integral extends over all field configurations $U(x,t)$ interpolating between $U(\vec{x},0) = U'(\vec{x})$ and $U(\vec{x},t) = U''(\vec{x})$. To obtain the analogous integral representation of the operator exp $(- tH)$, it suffices to replace the phase factor exp(iS) by exp$(- \tilde{S})$ where $\tilde{S}$ is the effective action in euclidean space

$$\langle U'' | e^{-tH} | U' \rangle = \int [dU] \, e^{- \int dx \, \widetilde{\mathcal{L}}_{eff}} \tag{4.2}$$

The integration extends over the same field configurations as in (4.1). The euclidean effective Lagrangian $\widetilde{\mathcal{L}}_{eff}$ is obtained from the quantity $\mathcal{L}_{eff}$ analyzed above by replacing the Minkowski space metric $g_{\mu\nu} = (+ ---)$ by $- \delta_{\mu\nu}$ and flipping the overall sign.

Finally, to arrive at the trace of exp$(- H/T)$, we take a time interval of length $1/T$, identify U'' with U' and integrate over U'. The functional integral now extends over all fields which are periodic in the euclidean time direction,

$$\text{Tr} \{ e^{-H/T} \} = \int [dU] \, e^{- \int dx \, \frac{\widetilde{\mathcal{L}}_{eff}}{T}}$$

$$U(\vec{x}, t + \frac{1}{T}) = U(\vec{x}, t) \tag{4.3}$$

Temperature thus produces remarkably little change: to obtain the partition function, rather than the vacuum transition amplitude, one simply restricts the manifold to a torus in euclidean space. The effective Lagrangian remains unaffected - the coupling constants F_0, B_0, ℓ_1, ℓ_2... are temperature independent. The essential point here is that the boundary conditions in the euclidean time direction are dictated by the trace which defines the partition function.

In the perturbative analysis of the partition function, the only change brought about by temperature is a modification of the pion propagator. In euclidean space, the massless propagator at zero temperature is

$$\Delta(x) = \frac{1}{4\pi^2(\vec{x}^2 + t^2)} \tag{4.4}$$

If the temperature is different from zero, the relevant propagator is the corresponding Green function on the torus,

$$G(x) = \sum_{n=-\infty}^{\infty} \Delta(\vec{x}, t + \frac{n}{T}) \tag{4.5}$$

The graphs associated with the partition function do not contain external legs. In the power counting rules, the role of the external momenta is taken over by the temperature, which is treated as a small quantity of order p. To calculate the partition function to a given order in the low temperature expansion, one needs to evaluate chiral perturbation theory to the corresponding order in the loop expansion.

In this language, the formula (2.6) represents the free energy density to one loop: the vacuum energy density ε_0 is the tree graph contribution, whereas the free gas term proportional to T^4 stems from the one-loop graph. Similarly, in (2.7), the first term represents the tree graph contribution and the term of order T^2 originates in the one-loop graph.

At first sight, it is surprising that, in the chiral limit, the low temperature expansion of the free energy density to order T^4 does not pick up a contribution from $L^{(4)}$. The reason why the constants ℓ_1 and ℓ_2 do not show up in (2.6) is that, at this order of chiral perturbation theory, $L^{(4)}$ only enters through tree graphs. The tree graph contribution to the logarithm of the partition function is given by the classical euclidean action, evaluated at the solution of the equation of motion, i.e. at U(x) = const. (If the quark mass vanishes, the solution is not unique, because the direction of the "magnetization" $<\bar{q}q>$ is arbitrary. For m > 0, the classical solution is given by U(x) = 1). The tree graphs are therefore temperature independent: at order p^4, $L^{(4)}$ only contributes to the vacuum energy (as can be seen from (3.3), this contribution in fact vanishes in the chiral limit).

More generally, in the evaluation of the free energy density to order p^n, $L^{(n)}$ only generates a temperature independent contribution to the vacuum energy. The term $L^{(n-2)}$ does give rise to a temperature dependent contribution, as it enters through one-loop graphs. As shown in[6], this contribution is however absorbed by pion mass renormalization. If the vacuum energy density is removed and if the remainder is expressed in terms of the physical pion mass, the expression for the free energy density to order p^n only involves the coupling constants of $L^{(2)}$, $L^{(4)}$,..., $L^{(n-4)}$.

5.   LOW TEMPERATURE EXPANSION OF THE CONDENSATE TO ORDER T^6

In ref. 1), the low temperature expansion of the partition function was worked out to two loops (z to order T^6, $<\bar{q}q>$ to order T^4). In the meantime, we have carried the low temperature expansion one step further by evaluating the perturbative expansion to three loops[6] (z to order T^8, $<\bar{q}q>$ to order T^6). I do not describe the details of the calculation here, but restrict myself to a brief discussion of the results obtained for the condensate. (The method used to evaluate the loop integrals us described in ref. 6); this reference also contains an analysis of the low temperature expansion for the pressure, for the energy density and for the heat capacity.)

In the chiral limit, the low temperature expansion is an ordinary power series in T/F and in logT and the evaluation of the graphs of chiral perturbation theory boils down to a calculation of the coefficients in this series, which are pure numbers. To order T^6, the result for the condensate reads[4]

$$<\bar{q}q> = <0|\bar{q}q|0>\left\{1 - x - \frac{1}{6}x^2 - \frac{16}{9}x^3\ln\frac{\Lambda_q}{T} + O(T^8)\right\}$$

$$x = \frac{T^2}{8F^2} \tag{5.1}$$

The formula shows that the temperature scale is set by $\sqrt{8}\,F \sim 250$ MeV. The origin of the term of order T^2 was discussed in section 2. Chiral symmetry fixes not only this term, but also the contribution of order

T^4 in terms of the pion decay constant. At order T^6, a logarithm appears. The symmetry determines the coefficient of the logarithm, but it does not determine its scale Λ_q. At this order in the low temperature expansion, the interaction generated by $L^{(4)}$ shows up: the scale Λ_q is determined by the coupling constants ℓ_1 and ℓ_2. The situation is analogous to the low energy expansion of the $\pi\pi$ scattering amplitude. As discussed in section 3, the low energy expansion of this amplitude to order p^2 only involves the pion decay constant, whereas, at order p^4, there is a contribution containing ℓ_1 and ℓ_2. It turns out that the particular combination of these two constants which fixes the scale Λ_q also specifies the leading term in the expansion of the isospin zero D-wave scattering length a_2^0 in powers of the quark mass. Eliminating the coupling constants, we obtain the following low energy theorem[4]:

$$a_2^0 = \frac{1}{144\,\pi^3 F^4}\left\{ \ln \frac{\Lambda_q}{M_\pi} - .067 + O(m)\right\} \quad (5.2)$$

(Note that in the limit $m \to 0$, the D-wave scattering lengths diverge logarithmically.) With this relation, the experimental information on a_2^0 allows us to determine the scale Λ_q phenomenologically[6]

$$\Lambda_q = 470 \pm 110 \text{ MeV} \quad (5.3)$$

The resulting parameter free low temperature representation of the condensate to order T^6 is shown in Fig. 1 (lower curve). As the temperature rises, the condensate gradually melts. At $T \sim 190$ MeV, the order parameter vanishes, indicating the occurrence of a phase transition at this temperature. Note, however, that the low temperature expansion cannot be trusted there: the curly bracket in (5.1) represents a correction. This is the case only in the upper part of the figure ($T \lesssim 150$ MeV). The error bars indicated in the figure only represent the sensitivity of the three-loop formula to the uncertainties in the constants F and Λ_q and do not account for the uncertainties due to higher order terms in the low temperature expansion or by contributions to the partition function generated by massive states (estimates for these are given in the next section).

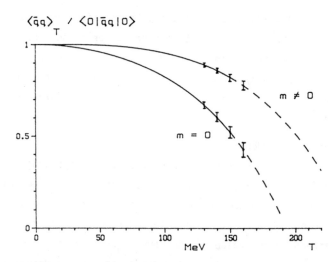

Fig. 1. Low temperature representation of the quark condensate to order T^6. The lower curve depicts formula (5.1), the upper curve is the result of a numerical evaluation for nonzero quark mass (M_π = 140 MeV). The error bars show the sensitivity of the prediction to the uncertainty in the value of the coupling constants.

For nonzero quark mass, the coefficents occurring in the low temperature expansion are not pure numbers, but are functions of the ratio M_π/T. At order T^2 and at order T^4, the coefficients can be expressed in terms of the kinematical function $h_1(M_\pi/T)$ introduced in section 2. At order T^6, there is a contribution from the three loop graph which we evaluate numerically. We again use phenomenological information to pin down the contributions generated by $L^{(4)}$. As a check on our calculations, we have verified that, at low temperatures ($T \ll M_\pi$), the result obtained for the pressure agrees with the familiar nonrelativistic formula which expresses the second virial coefficient in terms of phase shifts. The upper curve in Fig. 1 shows the numerical results for the condensate, evaluated at the physical value of the quark mass ($M_\pi \sim 140$ MeV). As was to be expected, it takes higher temperatures to melt the condensate if the pions are massive (the zero of the three-loop formula now occurs at T $\approx$ 240 MeV).

Although the u- and d-quark masses are tiny, their effect is quite substantial, because what counts in the Boltzmann factor is not the mass
of the quarks, but the mass of the pion, $M_\pi \sim \sqrt{m}$. In the chiral limit,
the condensate is reduced by almost a factor of two at a temperature
of the order of 150 MeV - this temperature is by no means large in
comparison to M_π.

6.   SIZE OF THE EXPONENTIALLY SMALL CONTRIBUTIONS

As discussed in the preceding sections, chiral perturbation
theory allows one to establish a number of exact statements concerning
the coefficients of the low temperature expansion. In this expansion,
massive states do not show up at any finite order, except indirectly,
through their effect on the value of the coupling constants. If we however leave the safe grounds of these low energy theorems and evaluate
the low temperature expansion at finite temperatures, then the contributions generated by the massive states cannot be ignored. I now turn
to an estimate of these contributions.

At low temperatures, massive particles are scarce, the density
being of order $n_i \sim \exp(- M_i/T)$, where M_i is the mass of the particle.
The interaction among the massive states manifests itself only at order
$n_i n_K \sim \exp -(M_i + M_K)/T$. Furthermore, chiral symmetry suppresses the
interactions between the massive particles and the pions by the factor
T^2/F^2. At low temperatures, the massive states can therefore be treated
as a free gas.

As shown in section 2, the contribution to the condensate
generated by a gas of free bosons of mass M_i is proportional to the
derivative $\partial M_i/\partial m$. To extend the calculation described there to all of
the known massive particles, it suffices to perform the sum

$$\Delta \langle \bar{q}q \rangle = \frac{1}{2} \sum_i g_i (M_i, T) \frac{\partial M_i^2}{\partial m} \tag{6.1}$$

over the entries of the table provided by the particle data group, accounting for the multiplicity of the levels with the appropriate spin
and isospin degeneracy factors[18].

The formula (6.1) makes sense only if the massive particles constitute a dilute gas. As the temperature rises, the density grows and it is then not legitimate to neglect their interactions. It is straightforward to calculate the density of the free particle data gas. One finds that, at a temperature of 140 MeV, the mean distance between two massive particles (K, η, ρ, N,...) is d = 2.2 fm. It rapidly shrinks as the temperature rises: at T = 160 MeV, we find d = 1.6 fm, whereas, at T = 200 MeV, the mean distance reaches d = 0.9 fm.

I conclude that the dilute gas approximation for the massive particles is meaningful up to a temperature of the order of 150 MeV. Beyond this point the model rapidly deteriorates.

To evaluate the contribution (6.1) we need an estimate for the value of the derivative $\partial M_i/\partial m$. In the nonrelativistic quark model, this derivative is equal to the number of u- and d-quarks contained in the particle. In the case of the K-meson, the σ-term $m \cdot M_K/\partial m$ is known rather accurately from the analysis of the pseudoscalar mass formulae. Since the nonrelativistic model does quite well for the kaon, we estimate the value of $\partial M_i/\partial m$ for all of the massive states on the basis of this model, using a fudge factor of two to cover our ignorance (for details, see ref. 6)).

At low temperatures, the contribution generated by the massive states is very small, less than 5 $^o/oo$ if T is below 100 MeV. Above T = 150 MeV, the contribution however rapidly grows. Even if the pions are ignored altogether, the dilute gas model predicts that the order parameter vanishes at T $\approx$ 200 MeV. Fig. 2a shows the effect of the massive states on the temperature dependence of the condensate in the chiral limit. The dash-dotted line depicts the three-loop formula (5.1). The shaded region represents the sum of the contributions generated by the pions and by the massive states. From this figure, I conclude that, in the chiral limit and up to a temperature of the order of 150 MeV, the melting of the condensate is mainly due to the pions, while the massive states only generate a small correction. The critical temperature is beyond the range of validity of our formulae. Since the order

74

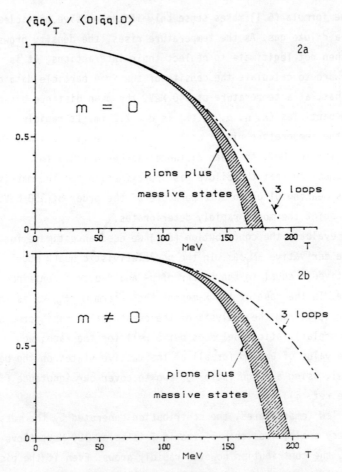

$\langle \bar{q}q \rangle_T / \langle 0|\bar{q}q|0 \rangle$

2a

m = 0

pions plus
massive states

3 loops

MeV

T

2b

m ≠ 0

3 loops

pions plus
massive states

MeV

T

Fig. 2. The dash-dotted lines represent the low temperature expansion to order T^6. The shaded areas result if the massive states are taken into account, using the dilute gas approximation.

parameter however rapidly falls at the upper end of this range, I nevertheless consider it meaningful to give an estimate. Fig.2a indicates
that, in the chiral limit, the phase transition occurs around

$$T_c \simeq 170 \text{ MeV} \qquad (m = 0) \qquad (6.2)$$

Fig. 2b shows the corresponding curves for nonzero quark mass
(M_π = 140 MeV). The figure indicates that the net effect of the quark
mass is to push the critical temperature up by about 20 MeV

$$T_c \simeq 190 \text{ MeV} \qquad (m \neq 0) \qquad (6.3)$$

In fact, the loop expansion still produces coherent results there (at T = 190 MeV, the one-loop formula predicts a reduction of the condensate by 27%; the two- and three-loop contributions increase this to 40 $\pm$ 5%.) The main uncertainty stems from the massive states. Even if their mutual interaction should turn out not to significantly affect their contribution to the condensate, at this temperature the pions are moving in a medium in which they encounter massive particles more often than they encounter pions. Accordingly, the estimate (6.3) should be taken with a grain of salt.

7.   CONCLUSION

The properties of the partition function can be analyzed in terms of a simultaneous expansion in powers of the temperature and in powers of the mass of the two lightest quarks (T $\rightarrow$ λT, m $\rightarrow$ λ^2m; expansion in powers of λ). The coefficients occurring in this low temperature expansion can be calculated by using the standard effective chiral Lagrangian. Chiral symmetry strongly constrains the structure of this Lagrangian and allows one to relate the coupling constants occurring therein to low energy observables such as the pion decay constant, the pion mass, the $\pi\pi$ scattering lengths etc. The first few coefficients of the low temperature expansion have been worked out explicitly. The available experimental information about the relevant low energy observables suffices to pin these coefficients down to within rather small uncertainties.

The limitations of the method are the following:
(i) Chiral symmetry determines the structure of the low temperature expansion only up to the values of the coupling constants. We have to rely on phenomenology to determine the values of these constants.
(ii) The expansion breaks down if the quarks are too heavy or if the temperature is too high.

In the analysis reported here, only the masses of the u and d quarks are treated perturbatively. Since these masses are very small, we expect the first few terms in the perturbative expansion to give a very accurate description. The limitation set by the requirement that the temperature must be low in order for the low temperature expansion to be useful is more serious. It limits the range where our results can be used quantitatively to $T \lesssim 150$ MeV. In particular, the low temperature expansion does not allow one to study the behaviour of the system in the vicinity of the critical temperature which we estimate at $T \approx 190$ MeV.

ACKNOWLEDGEMENT

It is a pleasure to thank the organizers of the conference for their kind hospitality. Furthermore, I wish to thank Ottilia Hänni and Peter Gerber for their help in the preparation of the manucript.

REFERENCES AND FOOTNOTES

1. J. Gasser and H. Leutwyler, Phys.Lett.B184, 83 (1987).

2. J. Gasser and H. Leutwyler, Phys.Lett.B188, 477 (1987).

3. H. Leutwyler, Phys.Lett.B189, 197 (1987).

4. For a review see
   H. Leutwyler, Nucl.Phys. B (Proc.Suppl.) 4, 248 (1988).

5. J. Gasser and H. Leutwyler, Nucl.Phys. B307, 763 (1988).

6. J. Gerber and H. Leutwyler, "Hadrons below the chiral phase transition", preprint University of Bern BUTP-25/88.

7. Y. Nambu and G. Jona-Lasinio, Phys.Rev. 122, 345 (1961); 124, 246 (1961).

8. J. Goldstone, Nuovo Cim. 19, 154 (1961).

9. M. Gell-Mann, R.J. Oakes and B. Renner, Phys.Rev. 175, 2195 (1968).

10. C.N. Yang and T.D. Lee, Phys.Rev. 87, 404 (1952).

11. In view of Monte Carlo simulations of the system which are necesarily performed at finite volume, it is of interest to analyze the finite size effects generated by the box. The method which I am describing in this talk is readily adapted to finite volume[5]; in fact, the leading finite size effects in some of the low energy observables have already been worked out[2,3,4].

12. A thorough discussion of the concepts underlying the notion of an effective Lagrangian can be found in refs. 13), 14). The extension of the method to nonzero temperature and to finite volume is discussed in detail in 5).

13. S. Weinberg, Phys.Rev.Lett. 18, 188, 507 (1967);
    S. Coleman et al., Phys.Rev. 177, 2239 (1969);
    C. Callan et al.. Phys.Rev. 177. 2247 (1969).

14. S. Weinberg. Physica A96. 327 (1979).

15. The Ward identities associated with $SU(2)_R \times SU(2)_L$ do not contain anomalies. For a detailed discussion of anomalies in the context of chiral perturbation theory, see
    J. Gasser and H. Leutwyler, Ann.Phys. 158, 142 (1984); Nucl.Phys. B250, 465 (1985).

16. The fields ϕ_1, ϕ_2, ϕ_3 are the canonical coordinates of the group element. One may choose any other parametrization of the group manifold; the results are independent of this choice.

17. H. Lehmann, Phys.Lett. B41, 529 (1972); Acta Physica Austriaca Suppl. 11, 139 (1973).

18. Since the temperatures of interest here are small compared to the mass of the proton, all of the particles can be treated as bosons.

O(3)-INVARIANT TUNNELING AT FALSE VACUUM DECAY

V.A.Berezin, V.A.Kuzmin, I.I.Tkachev

Institute for Nuclear Research of the
Academy of Sciences of the USSR,
Moscow, USSR

ABSTRACT

Tunneling processes at false vacuum decay in General
Relativity are studied. The general classification of
the bubble Euclidean trajectories is elaborated and
explicit expressions for bounces for some processes
like the bubble creation in the vicinity of a black
hole are given.

1. INTRODUCTION

The importance of phase transitions for the Universe
evolution was clearly understood in last years and a lot of
papers was devoted to study the subject (for reviews see
e.g. , [1,2]). The order parameter in these phenomena is the
magnitude of some scalar field φ ,fundamental or composite
(like $\overline{\varphi}\varphi$, G^2). The energy scale associated with the phase
transitions ranges from $\Lambda_{QCD} \sim 10^{-1}$ GeV corresponding to
the chiral symmetry breaking or to the quark confinement in
QCD through $M_X \sim 10^{15}$ GeV corresponding to Grand Unified
phase transitions up to $M_{Pl} \sim 10^{19}$ GeV which corresponds to
the possible compactification of extra space dimensions.

The basic object in investigations of phase transitions
phenomena is the effective potential V_{eff}. Its importance is
a reflection of the principial relation

$$V_{eff}(\varphi) = -p(\varphi) = \varepsilon(\varphi) - Ts(\varphi), \tag{1}$$

that is in a homogeneous medium V_{eff} is identical to the minus pressure within the system. In the relation (1) $\varepsilon(T,\varphi)$ is the energy as a function of temperature T and field φ , S being an entropy density. We learn that V_{eff} is the function of the temperature (Fig.1 shows an example of a typical behaviour).

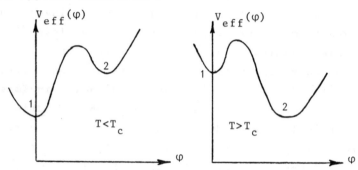

Fig.1. The scematic view of the effective potential below and above the critical temperature T_c.

Let the system be initially in the state 1. Fluctuations inevitably lead to nucleation of bubbles of the state 2. However, at $T > T_c$ the pressure in the state 2 (and so inside the virtual bubble) is smaller than in the surrounding phase 1. As a result the newly nucleated bubble collapses and the phase 1 remains stable.

In the course of the Universe expansion the temperature falls down and the effective potential takes the form shown in Fig.1b . Now the pressure in the phase 2 is smaller than in the phase 1 , the new phase bubbles expand and the phase transition proceeds.

We can read out also from relation (1) that $p=-\varepsilon$ at $T = 0$. This gives us the equation of state of a vacuum. Energy-momentum tensor components in a homogeneus isotropic

matter are
$$T_\alpha^\beta = \begin{pmatrix} \varepsilon & & \\ & -p & 0 \\ 0 & & -p \\ & & -p \end{pmatrix}$$

So, in a vacuum $T_\alpha^\beta = \varepsilon\delta_\alpha^\beta$, or $T_\alpha^\beta = \varepsilon\, g_\alpha^\beta$ which is the correct definition of a vacuum state even if we are working in the frameworks of General Relativity.

We should include gravity effects when considering cosmological or astrophysical phase transitions. Despite considerable interest in the problem of metastable vacuum decay in the early Universe, it still remains to be investigated properly even from the principal point of view. Investigation of bubble dynamics is an important part of this program. Here not only the motion of phase separation surface in real time is of interest but also its dynamics in Imaginary time which gives the probability of bubble creation should be perfectly analyzed. In the field theory one relates the probability p of subbarrier transition to the value B of a bounce, $p \sim \exp(-B)$, where $B = I_{new} - I_{old}$, I_{old} being the action (in imaginary time) for the solution describing the system in the initial (old) phase state while I_{new} in our case should be the action for the solution describing the system in the state with a bubble of a new phase. Generally speaking the validity of this formula for quantum gravitational transitions is not well founded for all cases, however, it is usually explored as well [3,4,5] .providing correct results even in some extreme cases [5]. We do not intend here to elaborate (or even to discuss) the foundations of this approach to the description of tunneling processes in quantum gravity but rather to analyse the structure of B in the framework of this approach for the case of O(3)-symmetric bubble solutions.

Considering bubble dynamics one usually uses the thin-wall approximation, i.e. phase separation surface is considered as infinitely thin and the whole problem is reduced to the one-dimensional (quantum) mechanics in this approach.

In the pioneering papers on this subject [6] there were not
taken into account effects of General Relativity, the calculations were carried out in a class of O(4)-invariant configurations. The first consideration of a vacuum decay in
the framework of General Relativity [3] was carried out also
for the O(4)-symmetric decay.

O(4)-invariance of the bubble means that the solution to
bubble equations of motion has a simple form $r = R \cos(\frac{\tau}{R})$
in imaginary time $(r = R \, ch(\frac{\tau}{R})$ for the real evolution),
where τ is a proper time on the phase separation surface,
R is the bubble radius at rest, or at the moment $\tau = 0$ of
its creation where both trajectories, real and imaginery,
join each other. This holds true even if we are working in
General Relativity framework but in the case of flat initial space-time O(4)-invariance means also that in Minkowsky
time t   the trajectory of a bubble wall has a form

$$r^2 - t^2 = R^2 . \qquad\qquad (2)$$

Coleman and DeLuccia [3] have assumed that the probability
of decay just to O(4)-symmetric configurations is maximal.
It was shown in our papers [7] that a homogeneous bubble of
a new phase does indeed possess of the O(4)-symmetry. However, the nucleation, for example, of a new phase bubble
with a remnant of the old phase inside the bubble is described by the O(3)-symmetric solution [7]. More general, O(4)-
invariant solutions appear only for the case when metrics
both inside and outside the bubble are the de Sitter one, so
the bubble nucleation say around the black hole should be
described by an O(3)-invariant trajectory.

Dealing with thin-wall approximation in the framework of
General Relativity it is the best way to use the Israel [8]
formalism of junction of two metrics on some singular hyper-
surface which is in fact Einstein equations for this surface.
A number of important results concerning thin-wall evolution

was obtained by Lake [9] . Directly to the vacuum phase transitions problems this formalism has been applied in
papers [7,10,11,12,4] . The trajectory of the bubble wall in
this approach is nothing but a shape of a three dimensional
phase separation surface. To define it one should first fix
the metric both inside and outside of the bubble and the
density of energy-momentum tensor on the shell. The latter
quantity for the pure vacuum bubble has the structure similar to that one induced by the Λ-term [7]

$$S_i^j = S\delta_i^j \quad , \tag{3}$$

where $i,j = 0,2,3$ are the tensor indeces on the hypersur-
face. The most general spherically symmetric vacuum metric
has a form

$$ds^2 = fdt^2 + \frac{1}{f}dr^2 + r^2 d\Omega^2 \quad , \tag{4}$$

where $f=1-\frac{2m}{rM_{Pl}^2} - \frac{8\pi r^2}{3M_{Pl}^2}\varepsilon + \frac{e^2}{r^2 M_{Pl}^2}$ and m,e,ε are Schwarzschild
mass, charge and vacuum energy density, respectively; all
three parameters do not in general equal zero in both regions. The Einstein equations on the shell give now $S = const$
and

$$4\pi SrM_{Pl}^{-2} = \sigma_{old}\sqrt{f_{old} - \dot{r}^2} - \sigma_{new}\sqrt{f_{new} - \dot{r}^2} \tag{5}$$

for the evolution of the bubble in imaginery time [7], where
$\dot{r} = dr/d\tau$. The sign functions σ are given by $\sigma_{old(new)} = +1$, if radii r of 2-dimensional spheres increase along the
direction of an outer normal to the 3-dim. phase separation
surface, and $\sigma_{old(new)} = -1$ in the opposite case. The derivative r has to be understood as the derivative with
respect to the proper time of an observer which is at rest
with respect to the bubble shell. So, only invariantly defined quantities come in Eq. (5) (the metric coefficients f

are functions of r only).

The Eq. (5) was investigated in papers [7,10,12]. It
is worth-while to note that in the limiting case $M_{P1} \to \infty$
this equation is nothing but that of energy conservation

$$m = -\frac{4\pi}{3}(\varepsilon_{old} - \varepsilon_{new})r^3 + 4\pi r^2 S \sqrt{1 + \dot{r}^2}. \quad (6)$$

It is easy to understand the meaning of all the terms   in
the right-hand side of Eq. (6). The first term describes a
potential energy of a charged shell; the second one is   the
difference between old and new vacuum energy densities; the
third is a, Wick rotated, kinetic energy of the shell.

One can treat   m   as a Hamiltonian of one dimensional
motion of a particle in a potential $W = m(r, \dot{r}=0)$. The form
of W which determines the motion of the bubble in real time
is shown in Fig.2.

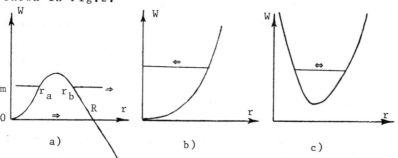

Fig.2. Schematic picture of W.   .a) $e = 0$, $\varepsilon_{new} < \varepsilon_{old}$,
     b) $e = 0$, $\varepsilon_{new} > \varepsilon_{old}$; c) $e = 0$, $\varepsilon_{new} > \varepsilon_{old}$.

We see that the uncharged remnant of the old phase (the
case of the Fig.2.b) inevitably collapses, but the charged
one can form bound states (Fig.2.c). New phase bubble with
$\varepsilon_{new} < \varepsilon_{old}$ can appear spontaneosly, moving first under
the   barrier in imaginery time, appearing at rest in turning
point R and then expanding in real time. Note, that due to

boundary conditions at infinity, the spontaneously created
bubble should appear with m=0. This immediately gives the
bubble radius at the moment of its creation

$$- \frac{4}{3}\pi R^3 \Delta\varepsilon + 4\pi R^2 S = 0, \qquad (7)$$

or

$$R = 3S/\Delta\varepsilon \quad .$$

However, one could create new phase bubble non-spontaneously
as well. For example, at studying physics on high energy
colliders (or due to cosmic rays) one could create subcriti-
cal bubble with $m \neq 0$ in the state of real motion at $r < r_a$.
That bubble could go through the burrier and then appear at
$r = r_b$ for the following infinite expansion. The probability
of subbarier tunneling from r_a to r_b should be larger than
from $r = 0$ to R (see Fig.2.a).

If the black hole is placed into metastable vacuum, its
mass enters the bubble equation of motion in the same manner
as m enters the Eq.(6) (since both are the Schwarzschild
parameters in the metric coefficients entering the Eq.(5)).

So, one may expect that black holes might initiate the
new phase bubble nucleation. However, in this case more
careful analysis is required (remind that in Eq(6) we took
the limit $M_{Pl} \to \infty$). In the present paper we obtain the
probability of new phase bubble nucleation in the vicinity
of a black hole with complete accounting for gravity effects
in the thin-wall approximation.

2. BOUNCE FOR O(3)-SYMMETRIC SOLUTIONS.

A solution to the Eq.(5) is O(4)-invariant only in the
case m=e=0. Here we consider bubble nucleation with a black
hole as a seed. One may consider a bubble to be O(3)-invari-
ant and correspondingly the total metric of space-time with
the bubble also to be O(3)-invariant (the same is true for

the metric before the bubble nucleation as well). A general
O(3)-invariant metric has a form

$$ds^2 = \gamma^{ij} dx_i dx_j - r^2 (d\theta^2 + \sin^2\theta d\varphi^2), \quad (8)$$

where θ and φ are angular variables; $i,j = 1,2$; $x_o = t$ is
the time coordinate and $x_1 = x_1(r)$ is the radial one. It was
shown in Ref. [13] that the following formula for the bounce
holds true

$$B = \frac{1}{2}\left[\int dx^2 \sqrt{\det\gamma}_{old} - \int dx^2 \sqrt{\det\gamma}_{new}\right] M_{pl}^2, \quad (9)$$

i.e. the bounce (and correspondingly the probability of bub-
le nucleation) is determined by the difference of areas of
two 2-dimensional surfaces θ = const, φ = const (the surface π for the further references) induced by (8) before
and after bubble nucleation. We would like to stress that
this formula has nothing to do with the thin-wall approximation, but of course it remains true in this approximation
as well. The expression (9) is extremely simple, elegant and
convenient for calculations of probabilities of any tunneling processes. One can make sure that this formula gives
correct results in all those cases where the value of B has
been found previously by means of straitforward calculations [3-6,14]. For example, at $M_{pl} \to \infty$ the purely geometric
expression (9) is reduced to the following well-known one [6]

$$B = \frac{\pi^2}{2} SR^3 \quad (10)$$

describing spontaneous new-phase bubble creation, where R
is given by (7).

3. GEOMETRY OF A VACUUM METRIC IN IMAGINARY TIME

According to (9), in order to calculate the bounce one needs to determine areas of 2-dimensional surfaces π's before and after the bubble nucleation. For simplicity, we consider here only one case of Schwarzschild world before bubble nucleation..In this Section we describe the form of the corresponding surface π before bubble creation; in the next Section we shall present the shape of π with the bubble. This might correspond to the real physical process initiated by a black hole provided the present state of the Universe with the vanishing vacuum energy density is metastable with respect to decay to the state with negative Λ-term.

We will be interested in the 2-dimensional surface with the metric

$$dl^2 = f\ dt^2 + f^{-1}\ dr^2 \qquad (11)$$

Since the metric coefficients do not depend on t, the surface to be found is axisymmetric and the time t is an angular variable. We introduce the dimensionless angle φ varying by 2π when going along the closed contour on the surface, then $t = T\ \varphi$ .The constant T is fixed by the requirement of absence of a cone singularity.For the Schwarzschild metric one has $f = 1 - 2m/rM_{Pl}^2$. In this case $T = 4m/M_{Pl}^2$. The corresponding surface π is drawn on Fig.3a. At large values of r the surface takes the form of a cylinder . The general form of the 3-dimensional section t = const for the Schwarzschild metric may be obtained by the rotation of the generatrix $\varphi = 0$ from Fig.3a. The result is shown in Fig.3b.

Spontaneous nucleation of a new phase bubble changes the shape of the surface π . We shall find it form and, correspondingly, its area in the thin-wall approximation , when

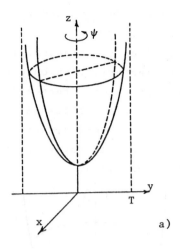 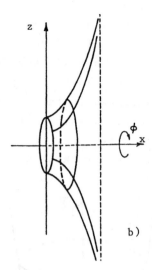

Fig.3. a) The surface π (11) for the Schwarzschild
metric. b) The 3-dimensional section t = const
of the Schwarzschild world.

by the definition, the solution with a bubble looks like the
initial surface but with a patch taken from the suface
(11) with $f = 1 - 2m/rM_{Pl}^2 - 8\pi\varepsilon_{new}/3M_{Pl}^2$, m and $\varepsilon_{new} < 0$ being
parameters of the metric inside the bubble. The equation of
the boundary of this patch is the equation of motion of bubble wall in imaginary time, and so is given by (5), and the
shape of the surface π which part is taken as a patch is
similar to that shown on Fig.3.a.

4. A CLASSIFICATION OF EUCLIDEAN TRAJECTORIES OF BUBBLE
   BOUNDARIES.

At r = 0 Eq.(5) has two roots which we shall denote by
r_a and r_b ($r_a < r_b$). The classical motion of the bubble
shell in real time may take place either in the interval
$0 \leqslant r \leqslant r_a$ or in the interval $r_b \leqslant r < \infty$. The Euclidean

trajectory of the shell oscillates in between r_a and r_b forming a closed contour on the surface π .

All such trajectories may be divided on two classes. To the first class we shall attribute trajectories, which could be shrinked to the point only by the extent of intersection of rotation axis of the surface. To the second class we shall attribute, correspondingly, the trajectories, which could be shrinked without intersection of the rotation axis. Trajectories of both classes are shown on Fig.4.

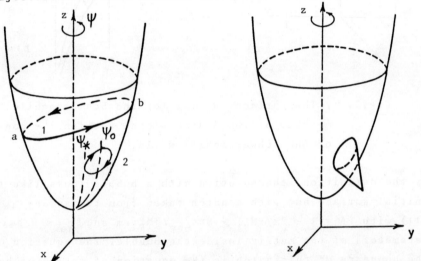

Fig.4.a)Trajectories of 1st (1) and 2nd(2) class on the
surface π(11); b) Surface π with a bubble.

Note, that this classification of trajectories should be given with respect to both the "new" metric and the "old" one. An example of the complete surface π with the bubble trajectory which is of the 2nd class with respect to the old metric (of the 2nd old class, for brevity) but of the 1st class in the new metric is shown on Fig.4.b

The trajectories of the types 1 and 2 describe entirely different physical processes. The section φ = const inter-sects a trajectory of the 1st class at any chosen value of φ , moreover this occurs only once. For trajectories of the

2nd class the situation is as follows. At $\varphi < \varphi_*$ the section φ = const does not contain the points of the trajectory at all, while at $\varphi_* < \varphi < \varphi_0$ it intersects the trajectory twice. The section φ_0 is distinguished by $(dr/d\varphi)|_{\varphi = \varphi_0} = 0$ both at the point a and b (see Fig.4.b). We give now the physical interpretation of both class trajectories [13].

1. The 1st old class trajectories describe the tunneling of a bubble from the state with $r = r_a$ to the state with $r = r_b$ (or vice versa) without change of metric parameters, and, may be, the bubble creation in a thermostat.as well.

2. The 2nd old class trajectories describe subbarrier creation of a new phase bubble (the shell b), which in turn contains inside a remnant of the old phase (the shell a). The shell a collapces in real time forming a black hole with a larger mass, while the shell b expands.

The quantitative criterion determining whether a shell belongs to the 1st or to the 2nd class was found in Ref.[13]. It could be easily expressed in terms of variables

$$ r_{\pm}^3 = -\frac{m_{new} - m_{old}}{8\pi^2 S^2 (\beta \pm 1)} M_{P1}^2 , \tag{12} $$

$$ \beta = \frac{\varepsilon_{new} - \varepsilon_{old}}{6\pi S^2} M_{P1}^2 . $$

At $f_{old} (r=r_+) > 0$ the trajectory is of the 2nd old class and it is of the 1st old class in the oposite case. At $f_{new}(r=r_-) > 0$ the trajectory is of the 2nd new class.

5. CALCULATION OF BOUNCES

The area of the part of the surface π bounded by the contour of bubble trajectory is $S = 2 \int_{\tau_a}^{\tau_t} dt dr$. Calculating it with respect to the old or to the new phase one has to use the corresponding time t . Using the relation between t and τ given by [10)]

$$\frac{dt}{d\tau} = \delta\frac{\sqrt{f - \dot{r}^2}}{f} \quad , \tag{13}$$

and equation of motion (5) one gets finally for the 2nd old class trajectories

$$S_{old}^{(2)} = -4\pi S(\beta + 1)M_{Pl}^{-2}\int_{r_a}^{r_b} dr(r_+ - r)(r_+^3 - r^3)r^{-2}/f_{old}|\dot{r}|, \tag{14}$$

The area of the surface π bounded by the 1st old class trajectory is

$$S_{old}^{(1)} = -4\pi S(\beta + 1)M_{Pl}^{-2}\int_{r_a}^{r_b} dr(r - r_{g,old})(r^3 - r_+^3)r^{-2}/f_{old}|\dot{r}| , \tag{15}$$

The formulas for the areas of surfaces π corresponding to new vacua restricted by the bubble trajectory may be obtained from Eqs (14) and (15) by the following substitutions

$$(old) \rightarrow (new), \quad r_+ \rightarrow r_-, \quad (\beta +1) \rightarrow (\beta -1).$$

In fact it is the black hole mass m_{new} which is a free parameter entering the definition of r_+ or r_- , and at given values of parameters of a theory the shell belongs to the 1st class or to 2nd one depending on the value of m_{new}.

We give finally the formulas for bounces for several particular processes.

1. *Vacuum creation of a concentric spherical ring of new phase with subsequent collapse of the inner shell down to the black hole.* One has

$$B = \frac{1}{2}(S_{old}^{(2)} - S_{new}^{(2)})M_{Pl}^2 \tag{16}$$

The Euclidean trajectory crosses both r_+ and r_-.

2. Subbarrier creation of the Einstein-Rosen bridge

Now the trajectory is of the 2nd class in the old metric
(therefore, it is the subbarrier tunneling indeed) but it is
of the 1st class in the new metric. The picture of the surface π_{new} is shown in Fig.4.b.,while the corresponding
Penrose diagram describing the evolution of such a world
in real time after the subbarier transition is drawn on
Fig.5. The bounce for this process is

$$B = \frac{1}{2}(S^{(2)}_{old} - S^{(1)}_{new})M^2_{Pl} .$$

(17)

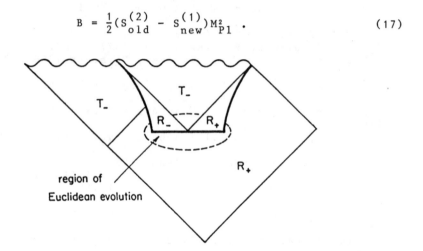

Fig.5. Subbarrier creation of the Einstein-Rosen bridge.

3. "Thermodynamical" creation of a new phase bubble.

The bounce is

$$B = \frac{1}{2}(S^{(1)}_{old} - S^{(1)}_{new})M^2_{Pl} .$$

(18)

We would like to note that our treatment corresponds presumably to a black hole in a thermostat since the temperature being defined as the period in imaginary time does not

depend on the distance from the black hole. It still remains
to be understood how to proceed with black holes in the
Unruh vacuum.

REFERENCES

1. A.D.Linde, Rep.Prog.Phys. 47 (1984) 925.
2. V.A.Kuzmin,M.E.Shaposhnikov,I.I.Tkachev, Sov.Sci.Rev.A
   Phys., Vol.8 (1987) 71.
3. S.Coleman,F.DeLuccia, Phys.Rev. D21 (1980) 3305.
   S.Parke, Phys.Lett. B121 (1983) 313.
4. W.Hiscock, Phys.Rev. D35 (1987) 1161.
5. S.W.Hawking, I.G.Moss, Ohys.Lett. 110B (1982) 35.
6. M.B.Voloshin,I.Yu.Kobzarev,L.B.Okun, Yad.Fiz. 20 (1974)
   1229.
   S.Coleman, Phys.Rev. D15 (1977) 2929.
   C.Callan, S.Coleman, Phys.Rev. D16 (1977) 1762.
7. V.A.Berezin,V.A.Kuzmin,I.I.Tkachev, Phys.Let.120B (1983)
   91.
8. W.Israel, Nuovo Cim. 44B (1966) 1;48B (1967) 463.
9. K.Lake, Phys.Rev.D19 (1979) 2847; D29 (1984) 1861.
10. V.A.Berezin,V.A.Kuzmin,I.I.Tkavhev, Phys.Rev. D36 (1987)
    2919.
11. V.A.Berezin,V.A.Kuzmin,I.I.Tkachev, Phys.Lett.124B (1983)
    479.
12. H.Sato, Prog.Theor.Phys. 76 (1986) 1250;
    P.Laguna-Castillo,R.Matzner, Phys.Rev. D34 (1986) 2913;
    K.Lake,R.Wevrick, Can.J.Phys. 64 (1986) 165;
    S.Blau,E.I.Guendelman, A.Guth, Phys.Rev. D35 (1987) 1747
    A.Aurilia,R.S.Kissack,R.Mann,E.Spallucci, Phys.Rev. D35
    (1987) 2961.
13. V.A.Berezin,V.A.Kuzmin,I.I.Tkachev, Phys.Lett. 207B (1988)
    397; preprint NBI-HE-87-83.
14. M.B.Voloshin,K.G.Selivanov, Pis'ma ZhETF 42 (1985) 342.

Thermal Hartree-Fock Calculations and Phase Transition in Nuclei

H. Sagawa

Department of Physics, University of Tokyo
Bunkyoku, Hongo 7-3-1, Tokyo 113
JAPAN

[An invited talk presented at Int. Summer Institute in Theoretical
Physics held at Queen's University, Kingston, Canada (July 25-29,1988)]

Abstract
    Static and dynamic thermal Hartree-Fock calculations are performed
in both nuclear matter and finite nucleus. We found that the critical
temperature of the liquid-vapor phase transition in nuclear matter is
proportional to the square root of the nuclear compressibility. A
microscopic model is applied to describe the time evolution of hot
nucleus ^{40}Ca. The model incorporates the dynamics of the expansion
process into finite-temperature Hartree-Fock theory. We found that most
of the nucleus remains in the liquid phase if the temperature is below 7
MeV, while most of it turns into vapor in the initial expansion for
higher temperatures.

1. Introduction and Summary

    Recently, intermediate energy heavy-ion collisions have been done
intensively, studying the validity of thermodynamical concepts in the
finite system.   One of the current topics is the liquid-vapor phase
transition in self-bound Fermi systems such as nuclear matter[1].
However, it is not clear that the critical point behavior of the phase
transition can be observed clearly in heavy-ion collisions since the
number of nucleons in the nucleus viewed as a many-body system is small.
It has been proved, nevertheless, that the critical temperature exists
for the formation of the compound nucleus[2]after the heavy-ion
collision.

    In this manuscript, I would like to report theoretical studies of
thermal nuclei using microscopic theories. First, formulas of Hartree-
Fock (H-F) theory are given in section 2 in the case of finite
temperature. This theory is particularly useful to calculate the
liquid-vapor phase transition since both phases come into naturally in
the calculations.   Some typical results of H-F calculations are

presented in section 3 using Skyrme density-dependent forces as the effective interaction. Density profiles of the closed shell nuclei ^{40}Ca and ^{208}Pb are reproduced very precisely by H-F calculations with the parameter set SGIII[3] which gives a low compressibility K_{NM} = 176 MeV.

The nuclear compressibility K_{NM} is a fundamental quantity in both nuclear matter and finite nuclei since the nuclear equation of state is basically governed by its value. Empirical information on the value K_{NM} was obtained through the excitation energies of isoscalar giant monopole states (so called the breathing modes), leading to $K_{NM} \approx$ 210-220 MeV [4],[5]. Recently, somewhat different values are adopted in the analysis of pion multiplicity from high-energy heavy-ion collisions[6] and the supernova explosion[7]. The stiff equation of state is prefered in the heavy-ion reactions, while the soft one is necessary for a successful explosion in a supernova simulation experiment.

We used three sets of the Skyrme force SIII[8], SGII[9] and SGIII in order to study the liquid-vapor phase transition in nuclear matter. The SIII force has a large compressibility K_{NM} = 356 MeV, while the SGII and SGIII forces show lower compressibilities K_{NM}= 216 MeV and 176 MeV, respectively, having lower powers of the density dependence. We find that the critical temperature of the phase transition is proportional to the square root of the compressibility

$$T_C \approx \sqrt{K_{NM}}$$

The mass asymmetry term of the nuclear compressibility is discussed using the analytic formula derived from the Skyrme force. We show that the SG III force having K_{NM}= 176 MeV gives a strong mass asymmetry dependence of the compressibility.

The time evolution of thermal nuclei is studied by using the double-constrained H-F hamiltonian in section 5. Quantum mechanical commutation relations are used to determine the time evolution. In the numerical test of our model for ^{40}Ca, it is shown that the temperature decreases rapidly during the initial expansion, while the entropy is kept constant. The theory predicts that there is a limiting temperature at around T= 7 MeV in the formation of compound nuclei. Experimental indications of this limiting temperature are reported in recent medium-energy heavy-ion experiments[2].

2. Hartree-Fock Approximation at Finite Temperature

I am going to study the dynamics governed by the many-body hamiltonian

$$H = \sum_{i,j} \langle i|t|j\rangle \, a_i^+ a_j + \frac{1}{2} \sum_{i,j,k,l} \langle ij|V|kl\rangle \, a_i^+ a_j^+ a_l a_k \qquad (2-1)$$

at finite temperature. The thermodynamical potential might be evaluated with a trial density operator

$$\hat{D} = \frac{1}{Z_G} \exp\{-\sum_i \alpha_i \, a_i^+ a_i\} \qquad (2-2)$$

where the factor Z_G is determined by the normalization condition $\mathrm{Tr}\{\hat{D}\}=1$. The trial parameters α_i will be determined by the variation procedure for the thermodynamical potential (the grand potential)

$$\Omega(T,\mu) = \langle H\rangle - T \cdot S - \mu\langle N\rangle \qquad (2-3)$$

$$= \mathrm{Tr}\{\hat{D}H\} - T*\mathrm{Tr}\{\hat{D}\ln\hat{D}\} - \mu*\mathrm{Tr}\{\hat{D}N\}$$

with respect to the density operator (the grand canonical partition function). The value μ represents the chemical potential. This variation procedure is an extension of the conventional Hartree-Fock (H-F) theory in which the hamiltonian $\langle H\rangle$ is minimized. Hereafter, the expectation value is determined by average ensemble with the density operator (2-2). The expectation value of the number operator is calculated to be

$$\mathrm{Tr}\{\hat{D}\,N\} = \mathrm{Tr}\{\hat{D}\,a^+a\} = \sum_i 1/\{1 + \exp(\alpha_i)\} = \sum_i f_i \qquad (2-4)$$

The expectation value of the hamiltonian is expressed by the probability function f_i as follows;

$$E = \langle H \rangle = \sum_i \langle i|t|i \rangle\, f_i + \frac{1}{2} \sum_{i,j} \langle ij|V|ij \rangle_a\, f_i\, f_j \quad (2\text{-}5)$$

where the two-body matrix element is anti-symmetrized. The entropy can also be expressed by f_i;

$$S = -\mathrm{Tr}\{\hat{D}\, \mathrm{tr}\,(\hat{D})\} = -\sum \{(1-f_i)\ln(1-f_i) + f_i \ln f_i\} \quad (2\text{-}6)$$

The thermodynamical potential is straightforwardly expressed as

$$\Omega = \sum_i \langle i|t|i \rangle\, f_i + \frac{1}{2} \sum_{i,j} \langle ij|V|ij \rangle_a\, f_i\, f_j$$

$$+ T \sum \{(1-f_i)\ln(1-f_i) + f_i \ln f_i\} - \mu \sum f_i \quad (2\text{-}7)$$

We will now derive the thermal H-F equations which will be derived by the variational procedures for the potential (2-8) with respect to the single-particle wave function and the probability function[10]. The minimization of the constrained potential

$$\tilde{\Omega} = \Omega - \sum \lambda_i \langle i|i \rangle \quad (2\text{-}8)$$

with respect to the single-particle wave function gives the H-F equation

$$\sum_i \langle i|t|i \rangle\, f_i + \sum_j \langle ij|V|ij \rangle_a\, f_j = e_i\, \delta_{ij} \quad (2\text{-}9)$$

where the single-particle energy e_i is defined by $e_i = \lambda_i / f_i$. On the other hand, the variation with respect to the probability function gives the formula

$$f_i = 1/[\, 1 + \exp\{(e_i - \mu)/T\}] \quad (2\text{-}10)$$

The H-F calculation at finite temperature is performed using the equation (2-9) with subsidiary conditions for the proton and neutron numbers,

$$\sum_{i\,\varepsilon\,\pi} f_i = Z \quad \text{and} \quad \sum_{i\,\varepsilon\,\nu} f_i = N \quad (2\text{-}11)$$

A temperature dependence of the H-F equation comes up with the density-dependence of the two-body interaction.

3. H-F calculations with Skyrme-interactions

The finite temperature H-F equations have two distinct solutions for given values of the chemical potentials μ_p and μ_n. One solution containing the liquid and vapor phases is obtained by starting the iteration procedure from a finite-range potential like a Woods-Saxon one. The second solution, on the other hand, having the vapor phase alone is obtained by starting the iteration from null potential[11]. The densities of the two solutions coincide asymptotically at large distance. In what follows we will denote the one-body density matrices by ρ^{L+V} and ρ^V which correspond to the first nucleus-like and the second vapor solutions, respectively,

$$\rho(\vec{r},\vec{r}') = \rho^{L+V}(\vec{r},\vec{r}') = \sum f_i \psi_i^*(\vec{r}) \psi_i(\vec{r}')$$

$$\rho^V(\vec{r},\vec{r}') = \sum f_i^V \psi_i^{V*}(\vec{r}) \psi_i^V(\vec{r}') \qquad (3\text{-}1)$$

where the probability functions f_i^V and f_i are calculated with the same chemical potentials $\mu^{L+V} = \mu^V$ for the liquid and vapor phases. Bonce et al., took the constraints for the proton and neutron numbers in the liquid phase

$$Z = \int \rho_p^L d\vec{r}, \qquad N = \int \rho_n^L d\vec{r} \qquad (3\text{-}2)$$

in order to obtain a good stability for the thermal H-F calculations. On the other hand, we will take different constraints in which the summed numbers both in liquid and vapor phases

$$Z = \int (\rho_p^L + \rho_p^V) d\vec{r}, \qquad N = \int (\rho_n^L + \rho_n^V) d\vec{r} \qquad (3\text{-}3)$$

are conserved. The latter choice might be more physical in the case of intermediate energy heavy-ion collisions since the liquid and vapor phases co-exist in the thermal nucleus.

We will now discuss numerical results of H-F calculations. As an effective interaction, the Skyrme-type density-dependent force is adopted. This interaction is parametrized as

$$V(\vec{r}_1,\vec{r}_2)=t_0(1+x_0P_\sigma)\delta(\vec{r}_1-\vec{r}_2)+t_1(1+x_1P_\sigma)(k^2+k'^2)/2*\delta(\vec{r}_1-\vec{r}_2)$$

$$+ t_2(1+x_2P_\sigma)\overleftarrow{k}'x\delta(\vec{r}_1-\vec{r}_2)\vec{k} + t_3(1+x_3P_\sigma)/6*\rho^\alpha*\delta(\vec{r}_1-\vec{r}_2) \qquad (3-4)$$

$$+ iW_0(\vec{\sigma}_1+\vec{\sigma}_2)\vec{k}'x\delta(\vec{r}_1-\vec{r}_2)\vec{k}$$

Table 1. Parameters of the Skyrme interactions and saturation properties of nuclear matter (fermi momentum k_F, binding energy per nucleon E/A, compression modulus K_{NM}, effective mass m^*/m and symmetry energy coefficient a_τ).

		SIII	SGII	SGIII
t_0	(MeV.fm^3)	-1128.75	-2645.0	-6243.2
t_1	(MeV.fm^5)	395.0	340.0	241.7
t_2	(MeV.fm^5)	-95.0	-41.9	116.4
t_3	(MeV.fm$^{3+3\alpha}$)	14000.0	15595.0	38071.0
x_0		0.45	0.09	0.06742
x_1		0.	-0.588	0.
x_2		0.	1.425	-2.807
x_3		1.	0.06044	0.
α		1.	1/6	1/15
W	(MeV.fm^3)	120.	105.	120.
k_F	(fm^{-1})	1.29	1.33	1.34
B/A	(MeV)	-15.9	-15.6	-15.6
K_{NM}	(MeV)	355.	216.	176.
m^*/m		0.76	0.79	1.
a_τ	(MeV)	28.2	26.8	28.0

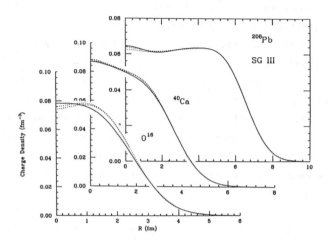

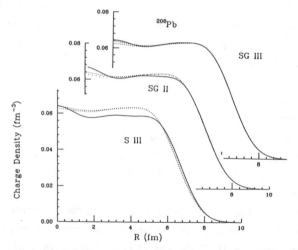

Fig. 1 Density profiles of charge distributions in ^{16}O, ^{40}Ca and ^{208}Pb. The solid curves show the calculated results, while the dashed ones are the experimental data with statistical errors. The SGIII force is used in the results shown in the upper part on the figure, while three different sets of the Skyrme dendity-dependent interactions are used in the calculations shown in the lower part. Experimental data are taken from ref.(12).

where P_σ is the spin exchange operator. The first three terms represent the S-wave, the P-wave momentum-dependent and the S- and D-wave momentum dependent interactions, respectively. The last two terms correspond to the density-dependent and the spin-orbit forces, respectively. Three sets of parameters used in this manuscript are tabulated in Table 1. It is known that the Skyrme force reproduces well the binding energies, the r.m.s. radii and the single-particle energies of many nuclei in a broad region of the mass table. As an example, the density profiles of some double-closed shell nuclei are shown in Fig. (1) together with experimental data[12]. The predictions obtained with the parameter set SGIII show excellent agreement with experimental data especially in the cases of ^{40}Ca and ^{208}Pb.

4. Nuclear compressibility and Critical Temperature of Liquid-Gas Phase Transition

The nuclear compressibility is a fundamental quantity of nuclear matter and also of finite nuclei. It is also the most important ingredient of the equation of state for studying the phase transitions. The nuclear compressibility has been discussed empirically in relation with the excitation energies of giant monopole resonances. Treiner et al.[5], obtained an empirical value K_{NM}= (220 +/- 30) MeV by assuming the scaling type transition density which exhausts 100 % of the sum rule value. The Saclay group[4] studied the giant monopole state using the random phase approximation (RPA) and also the Tassie-type transition density. They found the value K_{NM}= 210 MeV very close to that of Treiner et al..

Recently, the Groningen group[13] performed precise measurements of giant monopole resonances in Sm- and Sn-isotopes. Combining data of Sm- and Sn-isotopes with ^{208}Pb and ^{24}Mg (7 data points all together), they obtained the nuclear compressibility K_{NM} = 300 MeV performing the same analysis as Treiner et al.. This discrepancy between the Orsay and Groningen groups suggests that the analysis adopting a single collective state for the giant monopole resonance is ambiguous especially for light nuclei in which the width of the resonance is about 20 MeV. In order to obtain a reliable conclusion, we have to perform realistic RPA calculations in open-shell nuclei. So far, the RPA calculations of closed shell nuclei confirm the value of the compressibility K_{NM}= 220 MeV. It might be necessary to calculate GMR in open-shell nuclei in order to establish a solid relation between the value of K_{NM} and the excitation energy of GMR.

There are more arguments concerning the nuclear compressibility related with the analysis of pion multiplicity and multi-fragmentation after relativistic heavy-ion collision[6]. The Vlasov equation with the Uhling-Uhlenbeck collision term is used in the analysis of these data. A very stiff equation of states with $K_{NM} \gtrsim 400$ MeV is required to obtain a good agreement with the data. The equation of state is also crucial to study the Supernova explosion. Baron et al.[7] have successful explosions only when they used a soft equation of state having $K_{NM} \lesssim 180$ MeV.

We will study saturation properties of nuclear matter using the Skyrme force. The binding energy of the asymmetric nuclear matter is expressed analytically as follows;

$$E/A = 3\hbar^2 k_F^2/10m + 3t_0 \rho/8 + t_3 \rho^{\alpha+1}/16 + (3t_1 + t_2(5+4x_2))\rho k_F^2/80$$

$$+ \{ \hbar^2 k_F^2/6m - t_0(1/2 + x_0)\rho/4 - t_3(1+2x_3)\rho^{\alpha+1}/48$$

$$+ (3t_1 x_1 + t_2(4+5x_2))\rho k_F^2/24 \}(\rho_\tau/\rho)^2 \quad (4-1)$$

where $\rho_\tau = (\rho_n - \rho_p)$. The nuclear compressibility is defined as the second derivative of the binding energy with respect to the density at the saturation point;

$$K_{NM} = \rho^2 \frac{\delta^2(E/A)}{\delta \rho^2} \qquad at \ \frac{\delta(E/A)}{\delta \rho} = 0$$

$$= \{ -6\hbar^2 k_F^2/10m + 9t_3(1+\alpha)\alpha\rho^{\alpha+1}/16 + 3(3t_1 + t_2(5+4x_2))\rho k_F^2/8 \}$$

$$+ \{ 28\hbar^2 k_F^2/6m - 9t_0(1/2 + x_0)\rho/2 - t_3(1+2x_3)(1-\alpha)(2-\alpha)\rho^{\alpha+1}/48$$

$$+ (-3t_1 x_1 + t_2(4+5x_2))\rho k_F^2/24 \}(\rho_\tau/\rho)^2 \quad (4-2)$$

We take three different density-dependent interactions which have the compressibilities $K_{NM} \approx$ 180, 220 and 360 MeV, respectively, and study how the critical temperature is sensitive to the these values. The nuclear matter properties of three sets of the Skyrme-type interactions SIII, SGII and SGIII are listed in table 1. The essential difference among the three interactions is the power of the density dependence, namely, $\alpha=$ 1 for SIII, $\alpha=$ 1/6 for SGII and $\alpha=$ 1/15 for SGIII. The parameter set SGIII has been recently invented in order to simulate an interaction which has a low compressibility prefered in the supernova experiment. The calculated Hartree-Fock (H-F) properties of several nuclei with the SGIII interaction (the binding energies, the mean-square radii, the single-particle energies and the density profiles) will be published elsewhere.

The pressure is calculated from the first derivative of the free energy F = E - T*S with respect to the density,

$$P = \rho^2 \frac{\partial (F/A)}{\partial \rho} = P_{thermal} + P_{intrinsic} + P_{coulomb} \qquad (4\text{-}3)$$

where the pressure is divided into three parts; thermal, intrinsic and Coulomb pressure. The thermal part can be calculated from the kinetic energy at finite temperature. We used an analytic expression given by Jaqaman et al.[14],

$$P_{thermal} = \sum \alpha_n \, \rho^n = T*\rho + \alpha_2 (T)*\rho^2 + \text{----} \qquad (4\text{-}4)$$

The intrinsic part of the pressure can be obtained also analytically using eq. (4-1). The Coulomb part is discarded in the calculations shown in Fig. (2).

In Fig. (2), we show the results for the equation of states. In the region of high density, the liquid phase is stable when the curvature of the pressure is positive, while the vapor phase exists in the region of very low density. Comparing the three figures, it is interesting to notice that the interaction with higher compressibility gives larger negative pressure at the density around $\rho/\rho_0 \approx 0.7$ and therefore larger mixed phase region. On the other hand, the interaction SGIII having $K_{NM}=$ 176 MeV shows a smaller region of the mixed phase.

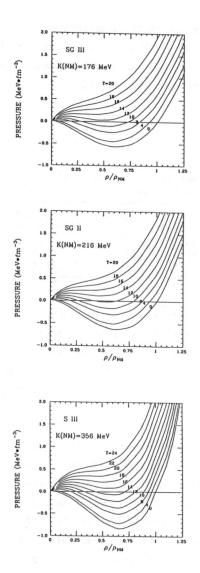

Fig. 2   Equation of states in the symmetric nuclear matter calculated
with the Skyrme forces at various temperatures.

We can extract the critical temperature T_C for the liquid-gas phase transition where $\partial P/\partial \rho = \partial^2 P/\partial^2 \rho = 0$ and the results are shown in Table 2. It is interesting that the critical temperature is proportional to the square root of the compressibility,

$$T_C \approx \sqrt{K_{NM}} \qquad (4-5)$$

Table 2. Nuclear compressibility and critical temperature of liquid-gas phase transition in nuclear matter

	K_{NM}(MeV)	T_C (MeV)
SGIII	176	14.4
SGII	216	15.8
S III	356	20.4

5. Time evolution of thermal H-F calculation

The time evolution of the reaction process has been studied using two different approaches. The first is the mean field theory like TDHF[15], while the second is based on the classical equations of motion[16]. Temperature is easily incorporated in the classical method, but there has been no serious calculation by TDHF at finite temperature, due to the numerical difficulty. Hereafter, I would like to present a hybrid model which was recently proposed by Bertsch and the present author[17]. A basic assumption of the model is that global thermal equilibrium is always existing during the expansion. This means that the proper time for the thermal equilibrium process is much shorter than that of the expansion process. While there is a certain limitation to apply this assumption to the high temperature case, it is worthwhile to pursue our model since there is no other study of the thermal system by microscopic quantum mechanical theory. Our final purpose is to calculate what fraction of the original nucleus will be left in the residual nucleus as a liquid phase after the expansion.

As was mentioned before, at finite temperature, the solutions of eqs. (2-9) and (2-10) have always components spreading over the entire coordinate space, besides those staying within the one-body potential similar to the solutions at zero temperature. Certainly, the former component which is identified as a vapor phase is sensitive to the size of the model coordinate space R_M as is shown in Fig.(3). Thus, the solutions might be physically meaningful in the following cases;

1) one looks at only the liquid solution putting a constraint on the liquid mass (ref.(11)).

2) The static solution with the vapor phase might be qualitatively meaningful when one calculates the equation of state from the constrained H-F results, because the equation of state is rather stable with respect to changes of the radius R_M (ref.(18)).

3) The solution with dynamical constraints might be relevant for describing the time-evolution of the thermal nucleus (ref.(17)).

4) The size of the spherical box is determined by other physical constraints, for example, a model of Wigner-Seitz cell for neutron star (ref. (19)).

Let me now describe the dynamical model. The dynamical constraints on the system might be determined by the physical situation as is described in the following. First, the nucleus might expand or shrink depending on the initial conditions. This situation is taken care of by a constraint term $\lambda_1 r^2$ (or equivalently $\lambda_1 \rho(r)$). Secondly, the nucleus might move collectively in a certain direction due to its expansion. We need another constraint term for this collective motion, namely, $\lambda_2 (\vec{p}.\vec{r}+\vec{r}.\vec{p})$. Thus, our model hamiltonian is described as a double-constrained one,

$$h"=h_{H-F}-\lambda_1 r^2 -\lambda_2 (\vec{p}.\vec{r}+\vec{r}.\vec{p}) \qquad (5-1)$$

where λ_1 and λ_2 are independent lagrange multipliers.

The dynamical equations of motion are written by using the quantum mechanical commutators. The time dependences of the expectation values of the operators r^2 and $\vec{p}.\vec{r}$ are given by

$$id\langle r^2\rangle/dt=\langle[r^2,h_{H-F}]\rangle=\langle[r^2,h"+\lambda_1 r^2+\lambda_2\vec{p}.\vec{r}]\rangle$$

$$id\langle\vec{p}.\vec{r}\rangle/dt=\langle[\vec{p}.\vec{r},h_{H-F}]\rangle=\langle[\vec{p}.\vec{r},h"+\lambda_1 r^2+\lambda_2\vec{p}.\vec{r}]\rangle \qquad (5-2)$$

where the expectation value implies averaging over the grand canonical ensemble,

$$\langle O \rangle = \sum_i f_i(T) \langle \Phi_i | O | \Phi_i \rangle$$

$$h'' \Phi_i(\lambda_1, \lambda_2) = \varepsilon_i(\lambda_1, \lambda_2) \Phi_i(\lambda_1, \lambda_2) \qquad (5\text{-}3)$$

Since the wave function Φ_i is an eigenstate of the constrained hamiltonian, we can rewrite eq.(5-2) as simple dynamical equations

$$d\langle r^2 \rangle / dt = 2\lambda_2 \langle r^2 \rangle, \qquad d\langle \vec{p} \cdot \vec{r} \rangle / dt = -2\lambda_1 \langle r^2 \rangle \qquad (5\text{-}4)$$

These coupled equations describe essentially a harmonic vibration mode (the breathing mode as the first sound) in the small amplitude limit. The macroscopic model by H. Schultz et al.[20] also gives a similar equation for the harmonic vibration.

6. The vapor-liquid phase transition in ^{40}Ca

The nucleus ^{40}Ca was taken for numerical study. We introduce 50 single-particle states as the neutron and proton configuration spaces, respectively. The vapor phase is obtained from one set of H-F solutions starting from zero potential with the same chemical potential μ as that of the solution with the vapor and liquid phases[11]. The second solution is calculated starting from the Woods-Saxon potential. Since we are going to study the ratio of the vapor to the liquid phases in the thermalized nucleus, the chemical potential is determined so as to keep the total number of particles in the liquid and the vapor phases fixed at A=40. We should notice that this prescription is different from that of Bonche et al.[7] who keep the number of particles fixed only in the liquid phase.

We now come to solve the time dependent equations (5-4) with physical initial conditions. As the first condition, we keep the excitation energy

$$E^* = \langle h_{H-F} \rangle^T_{\lambda_1, \lambda_2 = 0} - \langle h_{H-F} \rangle^{T=0}_{\lambda_1 = \lambda_2 = 0} \qquad (6\text{-}1)$$

to be constant. Secondly, it is assumed that the system when compressed at t=0 fm/c, has a normal density $\rho(r=0)\sim 0.15 fm^{-3}$ at the center. Namely, the lagrange multiplier λ_1 is taken to be negative, while the other multiplier λ_2 is zero at t=0 fm/c. There are no constraints for the temperature and the entropy during the time evolution. The equations (5-4) are solved with the time mesh $\Delta t=10$ fm/c and several different excitation energies (different initial temperatures T_I). The density profiles of the vapor and the liquid are shown in Fig.(3) at the beginning and the end of the first vibration with the temperature $T_I=6$ MeV on the l.h.s. and 12 MeV on the r.h.s.. Time dependence of temperature and entropy is shown in Fig. (4). The values λ_1, λ_2 and the

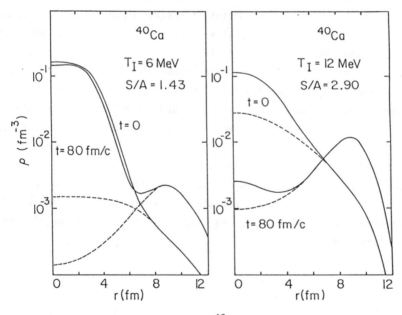

Fig. 3   The density distributions of ^{40}Ca at the initial and final stage of the expansion with the starting temperatures $T_I=$ 6 and 12 MeV, respectively.

temperature T are calculated at each time step to fulfill equations (5-4) and conservation of the total excitation energy. We can see that the temperature T is going down when the system is expanding. Nevertheless, the entropy S stays almost constant. Thus, in the initial stages of the expansion, the system behaves following the isentropic equation of state, but not the isothermal one.

In the case of T_I=6 MeV (equivalently, the excitation energy E^*=197 MeV and the entropy S/A=1.43), the vapor phase reaches 16% of the total nucleon number at the end of the expansion. On the other hand, the vapor phase occupies almost the entire space at t=80 fm/c in the case of T_I=12 MeV. This result implies that the whole system blows up at the very beginning of the expansion and never forms a compound-like nucleus at high temperatures. On the other hand, the system might oscillate until it cools down to a certain temperature, and expel some vapor particles isotropically (dominantly single nucleons) as is shown in Figs. (5) and (6). In Fig.(7), the ratio of the vapor to the liquid phases is drawn for different initial conditions. We can see that the vapor phase is less than 20% until the entropy S/A=1.5. The ratio of the vapor phase increases rapidly above this entropy reaching to almost 100% at the entropy S/A=2.9. Thus, there is a certain critical entropy S/A for the change of the dominant phase from liquid to vapor. Our model prediction is very close to that of the macroscopic model. Some empirical evidence on the formation of the compound nucleus has been reported in ref. (2) at a temperature of around T=6 MeV. Above this temperature, there is so far no empirical evidence of the compound nucleus formation.

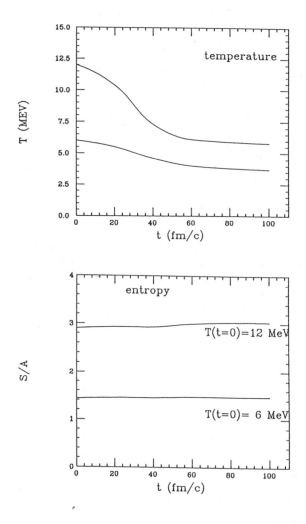

Fig. 4   Time   dependences   of the temperature and the entropy during the
expansion.

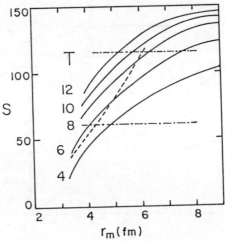

Fig. 5 The entropy S as a function of the radius. The dashed curve shows the result of Fermi gas model calculation at $T_I = 6$ MeV performing numerically the derivative of the thermodynamic potential with respect to the temperature. The dashed-dotted lines show the values obtained from $T_I = 6$ and 12 MeV.

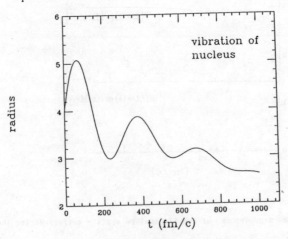

Fig. 6 Time evolution of the radius.

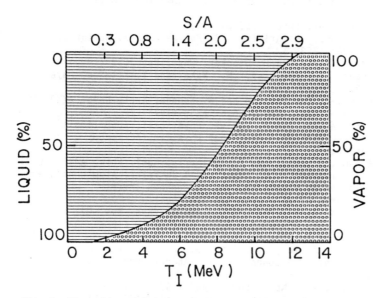

Fig. 7 The ratio of the mass number of the vapor phase to that of the liquid phase. The temperature T_I is that of the beginning of the expansion.

References

1) M. E. Fisher, Physics $\underline{3}$ (1967) 256
   C. B. Chitwood et al., Phys. Lett. $\underline{131B}$ (1983) 289
2) S. Song et al., Phys. Lett. $\underline{130B}$ (1983) 14
   J. P. Coffin, Proc.of RIKEN-IN2P3 Symp. on Heavy-Ion Collisions
   (Shimoda, Japan, Oct., 1987) p.304
   G. Guerreau, ibid., p.391
   J. Peter, ibid., p.357.
   R. Wada et al., Phys. Rev. Lett. $\underline{58}$ (1987) 1829
3) H. Sagawa, to be published
4) J. P. Blaizot, D. Gogny and B. Grammaticos, Nucl. Phys. $\underline{A265}$ (1976)
   315
   J. P. Blaizot, Phys. Rep. $\underline{64}$ (1980) 171
5) J. Treiner, H. Krivine . O. Bohigas and J. Martorell, Nucl. Phys.
   $\underline{A371}$ (1982) 253
6) J. J. Molitoris and H. Stocker, Phys. Rev. $\underline{C32}$ (1985) 346
   M. Sano, M. Gyulassy, M. Wakai and Y. Kitazoe, Phys. Lett. $\underline{156B}$
   (1985) 27
7) E. Baron, J. Cooperstein and S. Kahana, Nucl. Phys. $\underline{A440}$ (1985) 744
8) M. Beiner et al., Nucl. Phys. $\underline{A238}$ (1975) 29
9) Nguyen van Giai and H. Sagawa, Phys. Lett. $\underline{106B}$ (1981) 379
10) J. des Cloizeaux, "Many-body Physics" (edited by C. de Witt and R.
    Balian) p.5
11) P. Bonche, S. Levit and D. Vautherin, Nucl. Phys. $\underline{A427}$ (1984) 278
12) H. de Vries, C. W. de Jager and C de Vries, Atomic Data and Nuclear
    Data Tables $\underline{36}$ (1987) 495 and references therein
13) M. M. Sharma et al., preprint (KVI-691)
14) H. Jaqaman, A. Mekjian and L. Zamick, Phys. Rev. $\underline{C27}$ (1983) 2782
15) A. Dhar and S. Das Gupta, Phys. Lett. $\underline{137B}$ (1984) 303
    J. Knoll and B. Strack, Phys. Lett. $\underline{149B}$ (1984) 45
16) H. Jaqaman, A. Mekjian and L. Zamick, Phys. Rev. $\underline{C29}$ (1984) 2067
    A. Vicentini, G. Jacacci and V. Pandharipande, Phys. Rev. $\underline{C31}$ (1985)
    1783
    L. Vinet et al., Nucl. Phys. $\underline{A468}$ (1987) 321
17) H. Sagawa and G. F. Bertsch, Phys. Lett. $\underline{155B}$ (1985) 11
18) H. Sagawa and H. Toki, Prog. Theor. Phys. $\underline{76}$ (1987) 433
19) P. Bonche and D. Vautherin, Nucl. Phys. $\underline{A372}$ (1981) 496
20) H. Schultz et al., Phys. Lett. $\underline{147B}$ (1984) 17

QCD-Lattices and Sum Rules

GAUSS LAW AND SYMMETRY RESTORATION

A. Le Yaouanc, L. Oliver, O. Pène and J.C. Raynal

Laboratoire de Physique Théorique et Hautes Energies
Université de Paris XI, bât. 211, 91405 Orsay, France

M. Jarfi and O. Lazrak

Laboratoire de Physique Théorique, Faculté des Sciences de Rabat
Avenue Ibn Battouta, B.P. 1014, Rabat, Morocco

ABSTRACT

We study the restoration of global symmetries of lattice QCD at finite temperature and chemical potential for an arbitrary number of colors N_c and flavors N_f. The Hamiltonian in the A^0 gauge has to be supplemented by the Gauss law constraint, that thermal excitations must satisfy. We study the problem in the strong-coupling limit and in a Bogoliubov approximation. The free energy to be minimized must be defined by traces over states restricted in the color singlet Hilbert space at each lattice site. These restricted traces can be isolated by expressing general traces of a product of a unitary matrix U and a color invariant operator (like $\exp[-\beta(H-\mu N)]$ for the partition function) in terms of sums of products of characters of representations of $SU(N_c)$ and $GL(4N_f)$. We are able in this way to deduce the critical line for arbitrary N_c, N_f. The critical temperature increases with N_c due to Gauss law, but decreases with number of massless flavors N_f.

1. INTRODUCTION

There have been recently many studies of QCD at non zero temperature (T) and chemical potential (μ) using different techniques. The most powerful tool is the Monte-Carlo lattice calculation[1], but this technique has not yet been able to deal with non vanishing μ since the fermion determinant is no longer real. Perturbative QCD[2], as well as the effective chiral Lagrangian[3] have also been used. We will use[4] a strong coupling approximation in the Hamiltonian formalism. This is a drastic approximation which gives a not too realistic picture of reality since each hadron has all its constituents (quark or antiquarks) located at the same lattice site, without any glue excitation. Hence it is impossible to study the deconfinement transition. On the other hand the model is worth studying since at zero T and μ it gives dynamical breaking of chiral symmetry in a way that is qualitatively satisfying. Moreover this large g^2 limit is simple enough to be amenable to an analytical understanding of the phase transition including the case of non zero μ, for any N_c and N_f, with the use of some non trivial mathematics.

2. THE NAMBU-JONA-LASINIO GENERAL PICTURE

We do not mean here to use the Nambu Jona-Lasinio Lagrangian[5] which has recently attracted renewed interest, but the general philosophy developed by these authors.

2.1 Dynamical Chiral Symmetry Breaking

Consider any model of massless quarks subject to an attractive interaction in the color singlet sector. If the interaction is turned off, the physical Hilbert space is that of massless free quarks. When the interaction is strong enough to build up bound states (say scalar colorless bound states), these bound states, having a negative energy, destabilize the vacuum. The new vacuum contains a condensate of massless quark-antiquark scalar bound states, which interact with quark excitations. This interaction mixes positive and negative energy states leading to an energy gap between the positive and negative eigenvalues, i.e. to an effective mass. This is the constituent mass. The Hartree Fock approximation leads to a self-consistency condition between the generated mass and the condensate. This equation is usually called "gap equation" since it fixes the energy gap. To any solution of the gap equation which has a non vanishing condensate, corresponds massless solutions of the Bethe-Salpeter equation : the Goldstone bosons.

For a given chiral invariant Hamiltonian with massless quarks[5] (small quark mass) one can try to solve the gap equation, find the hadron spectrum, and, if chiral symmetry is dynamically broken, one gets massless (light) pions, and all the soft pion theorems derived from the (partial) conservation of the Axial Current.

2.2 Non Vanishing T and μ

Let us start from a solution of the gap equation as qualitatively described above. The physical vacuum, chiral non invariant, has a lower energy density than the would be chiral invariant vacuum. This is the mexican hat picture (Fig. 1). Now assume that we increase the baryon number density (equivalently μ). Each added quark adds to the system at least the energy gap, while the chiral invariant "vacuum" corresponds to massless quarks. Eventually, the chiral invariant solution becomes energetically favoured when the baryon density is large enough. This is the critical μ.

Fig. 1

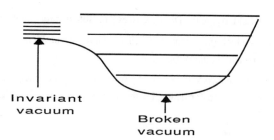

Invariant
vacuum

Broken
vacuum

For non-vanishing T, the thermal excitations amount to an average addition of quarks and anti-quarks with global fermion number equal to zero for $\mu = 0$. But here the picture is complicated by the presence of massless excitations in the chiral non invariant phase : the Goldstone bosons.

3. EFFECTIVE STRONG COUPLING HAMILTONIAN

Hamiltonian lattice QCD differs from the Lagrangian approach since it has a continuous time, while a discrete space. Obviously both should have the same continuum limit, but their strong coupling limits are quite different. Although the Lagrangian approach is more convenient for Monte Carlo simulation (path integral calculations), the Hamiltonian one has the advantage of exhibiting explicitly the structure of the Hilbert space.

3.1 The Kogut-Susskind Hamiltonian

We start from the following Hamiltonian[6] in the $A^0 = 0$ gauge.

$$H = H_G + H_D \tag{1}$$

$$H_G = \sum_{x,l,\alpha} \frac{1}{2a} g^2 E_l^\alpha(x) E_l^\alpha(x) + \sum_{x,l,m} \frac{1}{4g^2 a} \text{Tr} \{ 1 - U_l(x) U_m(x + a_l)$$

$$U_l^+(x + a_m) U_m^+(x) \} + \text{h.c.} \tag{2}$$

$$H_D = \sum_{x,l} \frac{1}{2ia} \psi^+(x) \alpha_l U_l(x) \psi(x + a_l) + \text{h.c.} \tag{3}$$

$$(l = 1...3)$$

where x labels the lattice sites, l,m the spatial directions, a the lattice spacing, ($\alpha = 1 ... 8$) are color indices, $E_l^\alpha(x)$ is the color electric field, and $U_l(x)$ are unitary matrices connected to the color magnetic fields, ψ is the quark field, α_l are the Dirac matrices $\beta\gamma_l$. Note that for simplicity we use here the naive Dirac fermions which suffer from the well known replica problem : there are $8N_f$ quarks. This could be cured by adding the Wilson term.

Due to the gauge invariance of the theory, the space of physical states is the space of the states invariant for time independent gauge transformation, i.e. the states verifying the Gauss law :

$$[\sum_l E_l^\alpha(x) + E_{-l}^\alpha(x) - \psi^+(x) \frac{\lambda^\alpha}{2} \psi(x) \ ] \ |\Psi> = 0 \ . \tag{4}$$

3.2 Effective Strong Coupling Hamiltonian

For N_c fixed we take the large g^2 limit[7]. The dominant contribution to (2) is

$$T = \sum_{x,l,\alpha} \frac{1}{2a} g^2 E_l^\alpha(x) E_l^\alpha(x) \ . \tag{5}$$

And the degenerate ground states verify

$$E_l^\alpha(x) \ |\Psi> = 0 \ . \tag{6}$$

Projecting out on the sector that verifies (6) and using second order perturbation theory one obtains the following effective Hamiltonian

$$H_{eff} = - \frac{K}{2 N_c} \frac{1}{a} \sum_{x,l} [\psi_a^+ (x) \alpha_l \psi_b(x + a_l) \psi_b^+(x + a_l) \alpha_l \psi_a(x)$$

$$+ (a \longleftrightarrow b, x \longleftrightarrow x + a_l)] \tag{7}$$

where

$$K = \frac{1}{g^2 \left(\frac{N_c^2 - 1}{2 N_c} \right)} \tag{8}$$

K sets the scale of hadron masses, of order $1/g^2$, while the glue excitations have energies of order g^2 as easily seen from (5). We restrict ourselves to the sector without glue excitations (6) which means that the gauge invariant states are necessarily built up of quarks and/or antiquarks all at the same site. The Gauss law reduces to

$$\psi^+(x) \frac{\lambda^\alpha}{2} \psi(x) \; |\Psi> = 0 \tag{9}$$

for any **x**.

4. DYNAMICAL SYMMETRY BREAKING

We first consider the case $T = \mu = 0$. Let us expand the quark field at $t = 0$

$$\psi(x) = \frac{1}{n^{3/2}} \sum_{k,s} [u_s(k) b_s(k) + v_s(k) d_s^+(-k)] e^{ik.x} \tag{10}$$

where u_s and v_s are trial Dirac spinors which define a trial vacuum (Bogoliubov states) through the equation

$$b_s(k) \; |\Omega> = d_s(k) \; |\Omega> = 0 \tag{11}$$

$|\Omega>$ are coherent superpositions of massless pairs.

Defining the projectors

$$\Lambda_+ = \sum_s u_s(k) u_s^+(k) \quad , \quad \Lambda_- = \sum_s v_s(k) v_s^+(k) \tag{12}$$

the vacuum energy $\mathcal{E} = <\Omega| H_{eff} |\Omega>$ is given by

$$\mathcal{E} = - \frac{K}{2N_c} \frac{1}{a} \frac{1}{n^3} \sum_{k,k'} 2N_c^2 \, \text{Tr} \, [\Lambda_-(k) \, \alpha_l \, \Lambda_+(k) \, \alpha_l] \quad . \tag{13}$$

If we assume rotational and P,C,T invariance of the vacuum, we can parametrize $\Lambda_\pm$ by

$$\Lambda_\pm(k) = \frac{1}{2} [1 \pm \sin \varphi(k) \, \beta \pm \cos \varphi(k) \, \alpha.k] \quad . \tag{14}$$

And $\varphi(k)$ is the new trial variable. The vacuum expectation value $<\psi\psi>$ is given by

$$<\psi\psi> = - \frac{2Nc}{n^3} \sum_k \sin \varphi(k) \tag{15}$$

$\sin \varphi(k)$ is proportional to a spontaneously generated mass, an effective mass, and it is not a

120

surprise that a non zero condensate implies a non zero effective mass.

The state which minimizes $\mathcal{E}$ is an extremum and must verify as a necessary (not sufficient) condition $\frac{\delta\mathcal{E}}{\delta\Lambda} = 0$, or equivalently

$$\frac{\delta\mathcal{E}}{\delta\varphi} = 0 \quad . \tag{16}$$

Equation (16) is the gap equation, often derived by other means. Once the solutions of (16) have been found, the approximate vacuum is the solution which has the lowest $\mathcal{E}$.

In the case of strong coupling QCD that we are considering, the gap equation (16) becomes

$$\sum_{k', \, l} [N_c \sin \varphi(k') \cos \varphi(k) - k_l \, k'_l \cos \varphi(k') \sin \varphi(k)] = 0 \quad . \tag{17}$$

Equation (17) has two solutions

$$\varphi = 0, \quad A = 0 \quad <\bar{\psi}\psi> = 0 \tag{18a}$$

$$\varphi = \frac{\pi}{2} \quad A = \frac{3K}{a} \quad <\bar{\psi}\psi> = -2N_c \, N_f \tag{18b}$$

where A is the generated quark mass. Solution (18a) is obviously a chiral symmetric one while (18b) corresponds to dynamical chiral symmetry breaking. It happens that (18b) has the lowest value $\mathcal{E}$:

$$\Delta\mathcal{E} = \mathcal{E}_b - \mathcal{E}_a = -\frac{K}{a} n^3 \, 3N_c \, N_f \sin^2 \varphi \tag{19}$$

5. FINITE T AND μ

5.1 Generalization of Bogoliubov approximation at $T \neq 0$ and $\mu = 0$

The free energy $F = \mathcal{E} - TS$ is given by

$$F(\rho) = \text{Tr} \, (\rho H) + \beta^{-1} \, \text{Tr} \, (\rho \log \rho) \tag{20}$$

in terms of the density matrix ρ. The minimization of $F(\rho)$ leads to the Gibbs formula :

$$\rho = Z^{-1} e^{-\beta H} \tag{21}$$

where Z is a normalization constant. In (21), H is the exact Hamiltonian, and it is very difficult to compute the traces Tr $(\rho \mathcal{O})$ for any operator $\mathcal{O}$. These traces (which lead to the thermodynamic averages) happen to be much simpler for Hamiltonians which are quadratic in terms of the quark fields (analogous to free Hamiltonians). Therefore the generalization of the Bogoliubov approximation consists in substituting to (21) a trial density matrix

$$\rho = Z^{-1} e^{-\beta H_2} \tag{22}$$

where H_2 is a trial quadratic Hamiltonian, in general β-dependent. Then it is easier to compute $F(\rho)$ and compute which ρ minimizes F.

Before implementing this program in the case of strong coupling lattice QCD, we must keep in mind the two following points

i) For H_{eff} (7) to be meaningful we must keep the conditions

$$T, \mu << \frac{g^2 N_c}{a} . \tag{23}$$

We will indeed find the critical values

$$T_c, \mu_c \sim O(\frac{1}{g^2 N_c a}) .$$

that satisfy (23).

ii) Thermal excitations must satisfy Gauss Law (equation (9)).

5.2 Calculation Forgetting (Momentarily) Gauss Law

To show how the things work, we will start by treating the simpler (but wrong) case when one forgets equation (9). This means that we take the trace in (20) over all states, including non color singlet states. We first assume $\mu = 0$. The partition function Z takes the simple factorized form :

$$Z = \det (1 + e^{-\beta H_2})$$

$$= \prod_{k,s} \left(1 + e^{-\beta E_s(k)}\right)^{N_c} \tag{24}$$

where $E_s(\mathbf{k})$ is the energy of one quark with momentum $\mathbf{k}$ and spin s :

$$H_2 = \sum_{\mathbf{k},s,\alpha} E_s(\mathbf{k}) \, b_{s,\alpha}{}^+(\mathbf{k}) \, b_{s,\alpha}(\mathbf{k}) \tag{25}$$

where α i s the color. We have assumed $N_f = 1$. We define

$$\Lambda(\mathbf{k}) = \sum_s n_s(\mathbf{k}) \, w_s(\mathbf{k}) \, w_s{}^+(\mathbf{k}) \tag{26}$$

where $n_s(\mathbf{k})$ is the average occupation number $(0 \le n_s(\mathbf{k}) \le 1)$ and $w_s(\mathbf{k})$ are the spinors (u or v):

$$\psi(\mathbf{x}) = \frac{1}{n^{3/2}} \sum_{\mathbf{k},s} w_s(\mathbf{k}) \, b_s(\mathbf{k}) \, e^{i\mathbf{k}.\mathbf{x}} \ . \tag{27}$$

From (26) one gets

$$\Lambda(\mathbf{k}) = \frac{e^{-\beta H(\mathbf{k})}}{1 + e^{-\beta H(\mathbf{k})}} \tag{28}$$

where $H(\mathbf{k}) = \sum E_s(\mathbf{k}) \, w_s(\mathbf{k}) \, w_s{}^+(\mathbf{k})$ (a Dirac matrix). The T = 0 limit is

$$n_+(\mathbf{k}) = 0 \quad \text{(positive energy states)}$$

$$n_-(\mathbf{k}) = 1 \quad \text{(negative energy states)}$$

$$\Lambda(\mathbf{k}) \xrightarrow[T\to 0]{} \Lambda_-(\mathbf{k}) \tag{29}$$

where $\Lambda_-(\mathbf{k})$ has been defined in (12). The gap equation is

$$\frac{\delta F(\Lambda(\mathbf{k}))}{\delta \Lambda(\mathbf{k})} = 0 \ . \tag{30}$$

If $\mu \ne 0$ we use the Grand Potential

$$\Omega = \mathcal{E} - TS - \mu <N> \tag{31}$$

where $N = \sum_{\mathbf{x}} \psi^+(\mathbf{x}) \, \psi(\mathbf{x})$.

Now Λ is given by

$$\Lambda(\mathbf{k}) = \frac{e^{-\beta(H(\mathbf{k}) - \mu)}}{1 + e^{-\beta(H(\mathbf{k})-\mu)}} \quad . \tag{32}$$

The gap equation

$$\frac{\delta\Omega(\Lambda)}{\delta\Lambda(\mathbf{k})} = 0 \tag{33}$$

leads to

$$H(\mathbf{k}) = -\frac{K}{a} \frac{1}{n^3} \sum_{\mathbf{k}',\mathbf{l}} \alpha_\mathbf{l}(1 - 2\Lambda(\mathbf{k}')) \alpha_\mathbf{l} \quad . \tag{34}$$

Parametrizing $H(\mathbf{k})$ by

$$H(\mathbf{k}) = A\gamma^0 + B\alpha.\mathbf{k} + C. \tag{35}$$

the gap equation becomes the set of coupled equations

$$A = m_0 \frac{sh(\beta A)}{ch(\beta A) + ch\beta(\mu-C)}$$

$$B = 0$$

$$C = m_0 \frac{sh(\beta A)}{ch\beta A + ch\beta(\mu-C)} \tag{36}$$

where m_0 is the solution for A at $T = 0$, $\mu = 0$ (eq. 18b) :

$$m_0 = \frac{3K}{a} \quad . \tag{37}$$

The order parameter is

$$<\bar{\psi}\psi> = -\frac{2N_c N_f}{a^3} \frac{A}{m_0} = -\frac{2N_c N_f}{a^3}(n_- - n_+) \tag{38}$$

and the fermion density is

$$\rho \equiv <\psi^+\psi> = \frac{2N_c\,N_f}{a^3}\,\frac{C}{m_0} = \frac{2N_c N_f}{a^3}(n_+ + n_- - 1) \tag{39}$$

in (38) and (39), $n_+(n_-)$ is the mean occupation number for positive (negative) energy states. The critical curve is found by the condition $A = 0$. This leads to

$$|\mu| = m_0\,\sqrt{1 - \frac{2T}{m_0}} + T\log\left(\frac{1 + \sqrt{1 - \dfrac{2T}{m_0}}}{1 - \sqrt{1 - \dfrac{2T}{m_c}}}\right). \tag{40}$$

The critical curve is plotted on Fig. 2. The critical temperature is for $\mu = 0$:

$$T_c = \frac{m_0}{2}\ . \tag{41}$$

For $\mu = 0$, $T < T_c$ the mean occupation numbers are

$$n_+ = 0,\quad n_- = 1\ . \tag{42}$$

For $T = T_c$ they suddenly jump to

$$n_+ = n_- = 1/2\ . \tag{43}$$

This means a total evaporation of the Dirac sea. Of course the change is so dramatic due to the crudeness of our model.

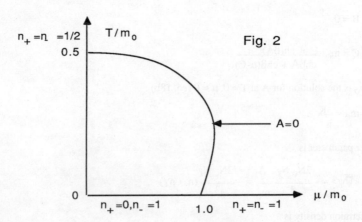

For T = 0,

$$\mu_c = m_0 \tag{44}$$

the occupation numbers are

$$n_+ = 0, \quad n_- = 1 \qquad \text{for } \mu < \mu_c$$

$$n_+ = n_- = 1 \qquad \text{for } \mu \geq \mu_c \ . \tag{45}$$

This means that for $\mu \geq \mu_c$ the density is the maximal allowed by Fermi statistics.

5.3 How to implement Gauss law ?

To implement Gauss law we must restrict the trace in (20) over the color singlet states, more precisely, from (9) the states which are singlet on each lattice site **x**. We will denote these restricted traces Tr_0 :

$$F_0(\rho) = \text{Tr}_0(\rho H) + \frac{1}{\beta} \text{Tr}_0(\rho \log \rho)$$

$$\text{Tr}_0(\rho) = 1 \ . \tag{46}$$

To show how to compute the restricted traces we will consider the simpler case of two colors. We use the orthogonality relations between the characters of the gauge group. Calling $\chi_T(\mathbf{a})$ the character of the group element $e^{i\mathbf{a}.\mathbf{T}}$ in the representation T :

$$\chi_T(\mathbf{a}) = \text{Tr}_T (e^{i\mathbf{a}.\mathbf{T}}) = \frac{\sin[(T + 1/2)|\mathbf{a}|]}{\sin(\frac{|\mathbf{a}|}{2})} \tag{47}$$

the orthogonality relation states that

$$\frac{1}{\pi} \int_0^{2\pi} da \sin^2(\tfrac{a}{2}) \, \text{Tr}_T(e^{iaT_3}) \, \text{Tr}_{T'}.(e^{iaT_3}) = \delta_{T,T'} \ . \tag{48}$$

For any operator the restricted trace on the representation T is given by

$$\text{Tr}_T(0) = \frac{(2T+1)}{\pi} \int_0^{2\pi} da \sin[(T + \tfrac{1}{2})a] \, \sin(\tfrac{a}{2}) \, \text{Tr}(0 \, e^{iaT_3}) \ . \tag{49}$$

For the singlet representation :

$$Tr_0(0) = \frac{1}{\pi} \int_0^{2\pi} da \, \sin^2(\frac{a}{2}) \, Tr \, (Oe^{\frac{iaT}{2}3}) \tag{50}$$

which we must apply at each lattice site. The Grand Potential (31) is given by

$$\Omega = Tr_0(\rho H_2) + \frac{1}{\beta} Tr_0 \, (\rho \, \log \rho) - \mu \, Tr_0 \, (\rho N) \quad . \tag{51}$$

The partition function is

$$Z_0 = \prod_x Z_0(x, X)$$

$$Z_0(x, X) = \frac{1}{\pi} \int_0^{2\pi} d\varphi(x) \, \sin^2 (\frac{\varphi(x)}{2}) \, Tr \, [e^{\psi+(x) \, -X+i\varphi(x) \frac{G_3}{2} \, \psi(x)}] \tag{52}$$

where $X = \beta(H-\mu)$. Mean fermion occupation numbers are computed from

$$< \sum_a \psi_{a\alpha}{}^+(x) \, \psi_{a\beta}(x)> = \frac{\partial \omega(X)}{\partial X_{\alpha\beta}}$$

$$\omega(X) = - \log Z_0 \quad . \tag{53}$$

The Grand Potential is given by

$$a^3 \Omega(\beta,\mu,X) = - \frac{K}{aN_c} \sum_1 tr \, [\omega'(X) \, \alpha_l(N_c - \omega'(X)) \, \alpha_l]$$

$$- \mu(tr \, \omega'(X) - 2N_c) + \frac{1}{\beta} [\omega(X) - tr \, (X \, \omega'(X)] \tag{54}$$

where ω' is the Dirac matrix defined in (53), and

$$\omega(X) = - \log Z_0(\lambda_i)$$

$$Z_0(\lambda_i) = \frac{1}{\pi} \int_0^{2\pi} d\varphi \, \sin^2\frac{\varphi}{2} \prod_i (1 + e^{-\lambda_i + i\varphi})(1 + e^{-\lambda_i + i\varphi}) \tag{55}$$

where the positive and negative energy eigenvalues of X, $\lambda_1 = \lambda_2 = \lambda_+$ and $\lambda_3 = \lambda_4 = \lambda_-$ are now our trial variables. The chiral transition happens when

$$\lambda_+ = \lambda_- \quad . \tag{56}$$

If we write

$$z_\pm = e^{-\lambda_\pm} \tag{57}$$

we see that the difference with the case where we neglected Gauss law is that, without Gauss law

$$Z = (1 + z_+)^{2N_c} (1 + z_-)^{2N_c} \tag{58}$$

while with Gauss law, Z is not factorised. Z is a more complicated symmetric polynomial, for example for $N_c = 2$:

$$Z = 1 + (3z_+^2 + 4z_+z_- + 3z_-^2) + (z_+^4 + 4z_+^3z_- + 10z_+^2 z_-^2 + 4z_+z_-^3 + z_-^4)$$
$$+ z_+^2 z_-^2 (3z_+^2 + 4z_+z_- + 3z_-^2) + z_+^4 z_-^4 \tag{59}$$

5.4 Critical curve for any N_c and N_f

We will simply state the final results refering the interested reader to ref. 4 for more details. The results for any N_c, $N_f = 1$ are plotted on Fig. 3. The critical chemical potential μ_c (at $T = 0$) does not change from its value m_0(eq. 44), while the critical temperature ($\mu = 0$) ($N_f = 1$) changes dramatically :

$$T_c = m_0 \frac{2}{15} \frac{(N_c + 1) (N_c + 2)(N_c+3)(N_c+4)(N_c+5)}{(N_c+1)(N_c+2)(N_c+3)(N_c+6) + 24} \xrightarrow[N_c \to]{} m_0 \frac{2}{15} N_c \quad . \tag{60}$$

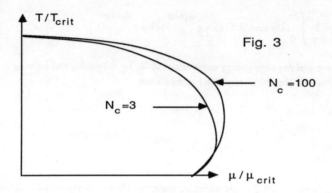

Fig. 3

The fact that T_c increases with N_c is no surprise : Gauss law restricts the number of states and thus a higher temperature is needed to excite enough physical states to wash out chiral symmetry breaking. If we now increase also the number of flavors we get

$$N_c \quad \to \quad : \quad T_c = m_0 \frac{2N_f}{16N_f{}^2 - 1} \left(N_c + \frac{4N_f{}^2 - 1}{N_f} + ...\right) \tag{61}$$

$$N_f \quad \to \quad : \quad T_c = \frac{1}{2} m_0 \left(1 + \frac{N_c{}^2 - 1}{2N_c} \frac{1}{4N_f} + ...\right) \quad . \tag{62}$$

We see that T_c decreases with increasing N_f, as expected from the fact that the number of physical states increases with N_f. Table I gives the value of T_c/m_0 for some values of N_c. Note however that the physical meaning of N_f is not clear since in the weak coupling limit the fermion replication gives a number of fermion species equal to $8N_f$.

Table I : T_c / m_0				
N_c	$N_f = 1$	$N_f = 2$	$N_f = 3$	$N_f = 20$
2	0.666	0.588	0.560	0.509
3	0.811	0.660	0.608	0.516
10	1.748	1.114	0.910	0.561
100	13.735	6.826	4.686	1.124

CONCLUSION

The main result is that in the limit of strong coupling Hamiltonian lattice QCD the Gauss law changes dramatically the "naive picture", it enhances the critical temperature which becomes proportional to N_c, but decreases when N_f increases. Of course, the model is only valid as long as the glue excitations remain frozen. This has prevented us from studying deconfinement. Furthermore, if the general trend of Gauss law to decrease the number of states is model independent, it may well happen that the glue degrees of freedom, allowing delocalized singlet states, would soften the N_c dependence of T_c. Finally, one could go beyong the Bogoliubov approximation to take into account the thermal excitations due to Goldstone bosons, which must be important at small T.

REFERENCES

1.   See for example : Karsch, F., Kogut, J.B., Sinclair, D.K. and Wyld, H.W., Phys. Letters B188, 353 (1987). Karsch, F. talk given at the 6th International Conference on "Ultrarelativistic Nucleus-Nucleus collisions", NordKirchen, West Germany, CERN-TH 4851/87. Kogut, J.B., Kovacs, E.V. and Sinclair, D.K., University of Illinois preprint, ILL-(TH)-87-29. Fukugita, M., Ohta, S., Oyanagi, Y. and Ukawa, A., KEK preprint 86-104.

2.   Redlich, K. and Turko, L., Z. Phys. C5, 201 (1980). Turko, L., Phys. Letters 104B, 153 (1981). Elze, H.-T., Miller, D.E. and Redlich, K., Phys. Rev. D35, 748 (1987). Müller, B., The Physics of the Quark-Gluon Plasma, Springer-Verlag (1985).

3.   Leutwyler, H., these proceedings.

4.   Le Yaouanc, A. et al., Phys. Rev. D37, 3691 and 3702 (1988) ; LPTHE Orsay 88/26.

5.   Nambu, Y. and Jona-Lasinio, G., Phys. Rev. 122, 345 (1961) ; 124, 246 (1961).

6.   Kogut, J. and Susskind, L., Phys. Rev. D11, 395 (1975).

7.   Smit, J., Nucl. Phys. B175, 307 (1980).

EXACT SELECTION RULE FOR 1^{-+} HYBRID DECAY INTO $\eta\pi$

F. Iddir, A. Le Yaouanc, L. Oliver, O. Pène, J.-C. Raynal

Laboratoire de Physique Théorique et Hautes Energies
Université de Paris-Sud, bâtiment 211, 91405 Orsay, France

ABSTRACT

We generalize in a model independent way a selection for 1^{-+} hybrid decay into $\eta\pi$ and $\eta'\pi$ previously derived in potential models, QCD sum rules and flux tube models. We show from QCD that only disconnected graphs contribute to these decays. It follows that these decays are OZI suppressed, with an additional suppression for $\eta\pi$ proportional to SU(3) breaking.

Since QCD is acknowledged as the theory of strong interactions, and has given rise to a qualitative understanding of the confinement mechanism, it is generally believed that no reason can prevent exotic hadrons to exist : glueballs (gg,...), hybrids ($q\bar{q}g$,...) multiquarks ($qq\bar{q}\bar{q}$,...). Many candidates for such exotics have been discussed, but it appers that is is usually very difficult to reach a clear unquestionable identification of such states. Of course, the situation would be better if one could observe a phanero-exotic hadron, i.e. a hadron whose quantum numbers cannot be accounted for by $q\bar{q}$ or qqq, such as $J^{PC} = 1^{-+}$, 0^{--} or 0^{+-} mesons.

Recently, a 1^{-+} meson decaying into $\eta\pi$ has been claimed to be seen by GAMS experiment[1] with a mass of 1400 MeV. It is therefore important to understand what can be predicted theoretically for a 1^{-+} meson.. In this note we concentrate on the hybrid ($q\bar{q}g$) interpretation. Several works[2] have studied the mass prediction for such states. The predictions for the lowest lying 1^{-+} hybrid lie from 1 GeV (QCD sum rules) to 1.9 GeV (bag model).

Fewer studies have been dedicated to the decay mechanism. It happens that the generalized quark model[3] (with constituent gluons), the flux tube model[4] and the QCD sum rules[5] predict a suppression of 1^{-+} ($q\bar{q}g$) $\to \eta\pi$. Finally this selection rule was generalized[6] in a model independent way from QCD.

We use as interpolating fields :

$$\pi^0 = \bar{u}\gamma_5 u - \bar{d}\gamma_5 d$$

$$\eta = \bar{u}\gamma_5 u + \bar{d}\gamma_5 d - \kappa\bar{s}\gamma_5 s$$

$$H^\mu = \left(T_{jk}^{\ i}\right)^\mu \bar{u}\left\{\gamma_i F_{jk}^{\ a}\frac{\lambda^a}{2}\right\} u - (u \leftrightarrow d) \tag{1}$$

$\left(T_{jk}^{\ i}\right)^\mu$ is any tensor combining the spatial vector indices i,j,k to build up a spin one object, namely the vector hybrid bound state $q\bar{q}g$. $F_{ij}^{\ a}$ is the gauge covariant color magnetic field. The isovector field H^μ has $J^{PC} = 1^{-+}$ as wanted for the interpolating field of the hybrid bound state.

The parameter k is chosen to decouple the η' ($\kappa \sim 1$ from experiment). In the following we shall work in the hybrid rest frame. We will also use and Euclidean metric. Let us now consider the following correlation function

$$F_\mu(x,y,z) = <H_\mu(z)\,\pi^0(x)\,\eta(y)> \quad . \tag{2}$$

We obtain

$$F_\mu(x,y,z) = F_\mu^I(x,y,z) + F_\mu^{II}(x,y,z) \tag{3}$$

$$F_\mu^I(x,y,z) = \frac{1}{N}\int \mathcal{D}(A)\,\delta(f(A))\,D\,\det\mathcal{M}_f\,e^{-S(A)}\left(T_{jk}^{\ i}\right)^\mu F_{jk}^{\ a}$$

$$x\left\{\mathrm{Tr}\left\{\gamma_i\frac{\lambda^a}{2}S_d^{(A)}(z,x)\,\gamma_5\,S_d^{(A)}(x,y)\,\gamma_5\,S_d^{(A)}(y,z)\right\}\right.$$

$$+\mathrm{Tr}\left\{\gamma_i\frac{\lambda^a}{2}S_d^{(A)}(z,y)\,\gamma_5\,S_d^{(A)}(y,x)\,\gamma_5\,S_d^{(A)}(x,z)\right\} + (u \leftrightarrow d) \tag{4}$$

$$F_\mu^{II}(x,y,z) = \frac{1}{N}\int \mathcal{D}(A)\,\delta(f(A))\,D\,\det\mathcal{M}_f\,e^{-S(A)}\left(T_{jk}^{\ i}\right)^\mu F_{jk}^{\ a}$$

$$x \left\{ \mathrm{Tr} \left\{ \gamma_i \frac{\lambda^a}{2} S_d^{(A)}(z,x) \gamma_5 S_d^{(A)}(x,z) \right\} + (d \leftrightarrow u) \right\}$$

$$x \left\{ \mathrm{Tr} \left\{ S_d^{(A)}(y,y) \gamma_5 \right\} + (d \to u) - \kappa(d \to s) \right\} \quad . \tag{5}$$

In the preceding formulae $\mathcal{D}(A)$ is for the path integration over the gauge fields, $\delta(f(A))$ is the gauge fixing term, $S(A)$ is the action, $\det \mathcal{M}_f$ is the Faddev-Popov determinant, D is the fermion determinant, $S_d^{(A)}(x,y)$ is the d-quark propagator in the A background field, and N is the normalization constant

$$N = \int \mathcal{D}(A) \, \delta(f(A)) \, \det \mathcal{M}_f \, e^{-S(A)} \quad . \tag{6}$$

In terms of Feynman graphs, $F^I(x,y,z)$ corresponds to the diagram of Fig. 1,

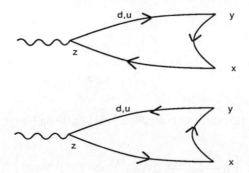

Fig. 1

Connected diagrams. The hybrid interpolating field is inserted in z and the $\eta(\pi)$ in y(x). The sum of a) and b) is zero.

while $F^{II}(x,y,z)$ corresponds to the diagram in Fig. 2.

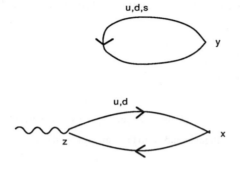

Fig. 2
Disconnected graphs. No cancellation appears here.

Let us define the Fourier transform.

$$\tilde{F}_\mu(p) = \quad dz\ dx\ dy\ e^{i(p.x - p.y)}\ F_\mu(z,x,y)\bigg|_{\substack{x_0 = y_0 = t \\ z_0 = 0}} \tag{7}$$

The decay amplitude of the hybrid H is given by $<H^\mu|T|\eta(-p)\ \pi^0(p)>$ with

$$\tilde{F}_\mu(p) \underset{t\to}{\to} C_H\ C_\eta\ C_\pi\ e^{-m_H t}\ <H_\mu|T|\eta(-p), \pi(p)> \tag{8}$$

where

$$C_H = \frac{1}{2m_H} <0|H_\mu(0)|H_\mu>$$

$$C_\eta = \frac{1}{2E_\eta} <0|\eta(0)|\eta(-p)>$$

$$C_\pi = \frac{1}{2E_\pi} <0|\pi^0(0)|\pi^0(p)>$$

$|p|$ has been chosen to be compatible with energy conservation in the decay of H ($E_\pi + E_\eta =$

M_H) and we have used the same notation for the states and their interpolating fields. The decay is P-wave and whence

$$F_\mu^I(\mathbf{p}) = - F_\mu^I(-\mathbf{p})$$

$$F_\mu^{II}(\mathbf{p}) = - F_\mu^{II}(-\mathbf{p}) \quad . \tag{10}$$

This is easy to check from (4) and (5) by inserting γ_0^2 in the traces, and comparing the contribution of a gauge field configuration $\mathbf{A}(x)$ to the Parity reversed one

$$\mathbf{A}'(x,t) = - \mathbf{A}(-x,t)$$

$$A'^0(x,t) = A^0(-x,t) \quad . \tag{11}$$

Next we notice from formula (4) that

$$F^I(z,x,y) = F^I(z,y,x) \tag{12}$$

leading to

$$\widetilde{F^I}(\mathbf{p}) = F^I(-\mathbf{p}) \tag{13}$$

which, combined with (10) gives

$$F^I(\mathbf{p}) = 0 \quad . \tag{14}$$

This is our main result : <u>only the disconnected diagram</u> (Fig. 2) <u>contributes to the 1^{-+} hybrid decay amplitude.</u>

To understand better how this result comes, let us consider an isosinglet 1^{-+} hybrid decaying into $\eta\eta$ (change in formula (1) the $-(u \leftrightarrow d)$ into a $+ (u \leftrightarrow d)$). The contribution of connected diagrams is now exactly the same as in eq. (4), while the disconnected ones contain, besides the result of eq. (5) a term analogous with x and y exchanged, leading to the vanishing of F^I and F^{II}. The total contribution vanishes although in this case this is no more than Bose statistics. The only difference, in the case under consideration of an isovector decay into $\eta\pi$ lies in the disconnected graph : the π, being isovector, is not coupled to the gluon sector, while the η is.

Now we would like to estimate the contribution of F^{II}. First, the contribution is proportional to $\sin \theta$ defined as

$$\eta = \eta_\theta \cos \theta - \eta_1 \sin \theta \quad . \tag{15}$$

In other words, for $\kappa = 2(\theta = 0)$ and $m_s = m_u = m_d$ it is obvious from (5) that F^{II} vanishes. Second, the disconnected diagrams of the type of Fig. 2 are $1/N_c$ suppressed as compared to the connected ones of Fig. 1. This is a hint for OZI suppression. Thus we conclude that the $\eta \pi$ decay amplitude of 1^{-+} hybrids is <u>OZI suppressed with an additional suppression factor of $\sin \theta$</u> (from 0.17 to 0.4).

A byproduct of our result concerns $\eta' \pi$ decay, which comes out, through the same line of argument, to be OZI suppressed but SU(3) allowed. Note however that the $1/N_c$ counting argument for OZI suppression may be misleading due to the important connection of η' and η to the gluon sector through the anomaly[7].

Finally notice that no suppression is found by the line of argument of this note for the $\pi\rho$ channel neither for $b_1(1235)\pi$ decay channel.

To conclude, we have shown in a quite model independent way, that the decay of any 1^{-+} hybrid into $\eta'\pi$ is OZI suppressed, and that the decay into $\eta\pi$ is further suppressed by $- \sin \theta$ ($\eta\eta'$ mixing angle).

REFERENCES

1. Binon, F., Proceeding of the Second International Conference on Hadron Spectroscopy, KEK, Tsukuba, Japan, p. 19 (April 1987). Gouanère, M., Proceedings of the International Europhysics Conference on High Energy Physics, Uppsala, Sweden, p. 564 (July 1987). Alde, D. et al. CERN-EP/88-15 (to appear in Phys. Letter B).

2. de Viron, F. and Weyers, J. , Nucl. Phys. B185, 400 (1981). Chanowitz, M. and Sharpe, S., Nucl. Phys. B222, 211 (1983) . Barnes, Y., Close, F.E. and de Viron, F., Nucl. Phys. B224, 241 (1983) .

3. Tanimoto, M., Phys. Letters 116B, 198 (1982). Le Yaouanc, A., Oliver, L., Pène, O., Raynal, J.-C. and Ono, S., Z. Phys. C, Particles and Fields 28, 309 (1985). Iddir, F., Le Yaouanc, A., Oliver, L., Pène, O., Raynal, J.-C. and Ono, S., Phys. Letters B205, 564 (1988).

4. Isgur, N., Kokoski, R. and Paton, J., Phys. Rev. Lett. 54 (1985) 869.

136

5. de Viron, F. and Govaerts, J., Phys. Rev. Letters $\underline{53}$, 2207 (1984).
6. Iddir et al., Phys. Letters $\underline{B207}$, 325 (1988).
7. Frère, J.-M. and Titard, S., University of Michigan preprint ; UM-TH-15 (1988).

QCD-Reactions and Deep Inelastic Processes

QUARK CORRELATION FUNCTIONS
AND DEEP INELASTIC SCATTERING

C.H. Llewellyn Smith,

Department of Theoretical Physics,
1 Keble Road,
Oxford,
OX1 3NP.

ABSTRACT   It is shown that, when ultraviolet divergences are properly taken into account, the naive relation between deep inelastic structure functions and quark light-cone correlation functions is replaced by a relation to quark correlation functions smeared over distances of order $1/\sqrt{Q^2}$ transverse to the light-cone. The quark and gluon distributions, which - with the definition adopted here - are related to the structure functions by a convolution, are shown to be Fourier transforms of the smeared correlation functions to all orders in α_s but zeroth order in $1/Q^2$. The discussion is based on a version of the all orders in α_s proof of factorization of mass singularities which leads to a particularly transparent proof of the evolution equations. The relation to correlation functions allows structure functions to be predicted in models and underwrites quantitative discussions of the distances that control deep inelastic scattering.

1) INTRODUCTION

It was recognized almost twenty years ago that deep inelastic lepton scattering measures the commutator of two currents near the light-cone [1] and that if the light-cone singularity is the same as in the free quark model, the large Q^2 behaviour of structure functions is determined by the quark light-cone correlation function [2]. It was also recognized from the outset that the relation to the light-cone correlation function is a purely formal one which suffers from divergences in renormalized field theory [3]. This relation has nevertheless been used recently as the basis for model calculations of structure functions [4] and as a tool for determining the distances which play an important role in deep inelastic scattering [5], but the significance of the results is obscure as the relation is purely formal. In this paper I show that in interacting field theories structure functions are related to quark correlation functions smeared over distances of order $1/\sqrt{Q^2}$ transverse to the light-cone and derive a precise relation between this quantity and the deep inelastic structure functions which holds to zeroth order in $1/Q^2$ (or leading twist if quark masses are neglected) and all orders in the strong coupling constant*. The fact that the dominant distances in deep inelastic scattering are within $1/\sqrt{Q^2}$ of the light-cone is well known but the precise relationship between structure functions and the smeared correlation function has never been given in the literature to the best of my knowledge.

Section 2 contains a review of the naive relationship between deep inelastic scattering and light-cone correlation functions obtained by assuming that light-cone singularities have the same structure as in free-field theory. Turning to interacting field theory in section 3, we review the results obtained using the operator product expansion (OPE) and relate the structure functions to a formal light-cone correlation function defined in terms of the local quark-bilinear operators that enter the OPE. The diagrammatic study of the factorization of mass singularities in the leading twist contributions to structure functions is used in section 4 to show that the results in sections 2 and 3 can be given a well-defined meaning if the formal light-cone correlation function is replaced by a correlation function smeared over distances of order $1/\sqrt{Q^2}$ transverse to the light-cone. The discussion of factorization is essentially equivalent to the treatments in the literature, although the specific form of projection operator used to achieve factorization makes the analysis particularly simple. It seems, however, that the fact that the "soft" parts of the factorized amplitudes (which we take to be equal to the quark and gluon distributions by definition) are related to smeared correlation functions has not been pointed out explicitly before, and the derivation of the evolution (renormalization group) equations for the distribution functions also seems new and particularly transparent. Section 5 contains remarks on model calculations of structure functions and on the dominant distances in deep inelastic scattering on nuclei and nucleons.

2) FREE-FIELD RESULTS

Deep inelastic cross-sections are related to the imaginary part of the forward virtual Compton amplitude by the optical theorem, as depicted in Fig 1. The cross-section

---

* In [5] I referred (in rather misleading terms) to a "renormalized" light-cone correlation function, by which I meant the smeared function, and referred to a forthcoming paper for details; that paper (which would have presented the zeroth order in α_s case of the treatment given here) was unfortunately never completed. Collins and Soper [6] introduce a renormalized light-cone correlation function, but it has the same status as Eq (13) below i.e. its moments are renormalized separately, or - equivalently - it is related to the unrenormalized cut-off dependent correlation function by a convolution, and does not have a simple interpretation.

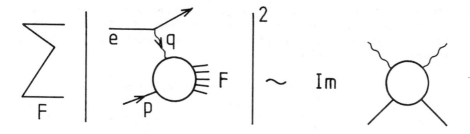

Fig 1   Inclusive deep inelastic lepton scattering.

is given by

$$\sigma \sim \epsilon_\mu^* \epsilon_\nu \int e^{iq\cdot y} < p|J_\mu^\dagger(y)J_\nu(0)|p >_c d^4y$$

where ϵ is the polarization vector of the virtual photon or vector boson and c denotes a connected matrix element. With the standard variables $\nu = q.p$, $Q^2 = \vec{q}^{\,2} - q_0^2$ and $x = Q^2/(2\nu)$, we have

$$q.y = \frac{\nu y_-}{M} + \frac{Mx}{2}(y_+ + y_-),$$

in the target rest frame, in terms of light-cone variables $y_\pm = y_0 \pm y_3$ where the 3 axis is defined to be along $\vec{q}$. Thus for large ν

$$\sigma \sim \int dy_- e^{i\nu y_-/M} \int dy_+ \, d^2y_T \, e^{iM \, xy_+/2} \, \epsilon_\mu^* \epsilon_\nu < p|J_\mu^\dagger(y)J_\nu(0)|p >_c .$$

In the Bjorken limit $\nu \to \infty$ with x fixed, the rapidly oscillating exponential factor kills all contributions except those from regions where the integrand is singular. Since singularities are only expected for $y^2 = 0$ (except if $x = 0$, i.e. for photoproduction, in which case the y_+ integral is not damped by an oscillating factor and can diverge and produce other singularities), deep inelastic scattering is light-cone dominated. The matrix element depends only on y_0 and $|\vec{y}|$ so the angular integral is easily done yielding phase factors $\nu(y_0 \pm |\vec{y}|)/M \pm Mx|\vec{y}|$. It then follows from the fact that the matrix element is singular only at $y_0 = \pm|\vec{y}|$ and decreases with $|\vec{y}|$ along the light-cone that the dominant regions are $|y_0 \pm |\vec{y}|| \lesssim M/\nu, |\vec{y}| \lesssim 1/Mx$.

It is instructive to consider what would happen if the light-cone singularities in $J_\mu^\dagger(y)J_\nu(0)$ were the same as in free field theory i.e. equal to a singular C number function of y times a quark bilinear evaluated at $y^2 = 0$, apart from a C number term which does not contribute to connected matrix elements. Since the degree of singularity is given by ordinary dimensional analysis in free field theory, there would be exact Bjorken scaling and a parton model description would hold with parton distributions given by

$$q_i(x) = \frac{1}{2\pi} \int e^{-ip_+ zz} < p|\bar{\Psi}_i(\hat{z})\gamma_+ \Psi_i(0)|p >_c dz$$

$$\bar{q}_i(x) = \frac{-1}{2\pi} \int e^{ip_+ zz} < p|\bar{\Psi}_i(\hat{z})\gamma_+ \Psi_i(0)|p >_c dz$$

$$(1)$$

where $\hat{z}_\mu = (z,0,0,-z), \gamma_+ = \gamma_0 + \gamma_3, p_+ = p_0 + p_3$ and i is a flavour label. Eq (1) is rendered meaningless by ultraviolet divergences in interacting field theory but, since it does exist in models endowed with a cut-off and it is not dissimilar to the smeared quantity which replaces it in interacting theories, it is worth pursuing its formal properties a little further.

Defining $\Psi_+ \equiv \gamma_0\gamma_+ \Psi/2$, we note that $\bar{\Psi}\gamma_+ \Psi = 2\Psi_+^\dagger \Psi_+$ and recall that if the theory is quantized on the light-cone the fields Ψ_+ and $\Psi_+^\dagger$ satisfy canonical equal $z_+(= z_0 + z_3)$ anti-commutation relations. Consequently we can replace $\Psi_+^\dagger(\hat{z})\Psi_+(0)$ by $-\Psi_+(0)\Psi_+^\dagger(\hat{z})$ if we wish since the C number anti-commutator does not contribute to connected matrix elements. Eqs (1) can therefore be rewritten in the obviously parton-like form *

$$q_i(x) = 2\sum_n \delta\left(p_+^n - (1-x)p_+\right)| < n|\Psi_{i+}(0)|p > |^2$$

$$\bar{q}_i(x) = 2\sum_n \delta\left(p_+^n - (1-x)p_+\right)| < n|\Psi_{i+}^\dagger(0)|p > |^2$$

$$(2)$$

from which it follows that q and $\bar{q}$ are both zero in the non-physical region $x > 1$. It is easy to check that only the fully connected pieces of the matrix elements in (2) are non-vanishing for $x > 0$. The distributions in the non-physical region $x < 0$

---

* An analysis based on the Feynman diagrams which contribute to deep inelastic scattering (as in section 4) leads naturally to (2), ignoring ultra-violet divergences, whereas (1) is generated by consideration of the light-cone singularity. Jaffe [7] starts from an analysis of the T product $T(\Psi_+^\dagger(\hat{z})\Psi_+(0))$, which appears rather unnatural in the calculation of a cross-section but is equivalent to an ordinary product (ignoring the fact that neither is well defined) as far as connected matrix elements are concerned (although his proof involves several illegitimate steps and his claim, in mitigation, that the relevant matrix elements are two particle irreducible, and that therefore his proof is legitimate and Eq (1) is ultraviolet convergent, is wrong).

are related to the physical distribution by the crossing relation $q_i(x) = -\bar{q}_i(-x)$ which follows directly from (1).

We now recast (1) in a form which shows immediately that it has no meaning, except in models with an ultraviolet cut-off, and also allows us to make contact with the operator product expansion. Inverting (1) and making a power series expansion, we find

$$\int_0^1 x^n \left(q_i(x) \mp \bar{q}_i(x)\right) dx = \frac{i^n}{(p_+)^{n+1}} < p | \bar{\Psi}_i(0) \gamma_+ \left(\partial_+\right)^n \Psi_i(0) | p >_c \qquad (3)$$

where the $-(+)$ sign applies to the case n even (odd). The matrix elements of the operators on the right hand side of this equation are all ultra-violet divergent (except for $n = 0$ and Ψ unrenormalized); they can only be rendered finite if each operator is interpreted as a Zimmermann normal product [8] and renormalized, in an n dependent manner.

3) THE OPERATOR PRODUCT EXPANSION

The operator product expansion (OPE) [9, 8, 10] controls the $q_0 \to i\infty$ behaviour of the amplitudes for forward scattering of virtual photons and vector bosons which are related to structure functions by dispersion relations. The result has the form [11]

$$\int x^n F^a(x, Q^2) dx = \sum_i \hat{C}_i^{a,n}(Q^2, \mu^2, m^2) < p | 0_i^n(\mu^2) | p >_c \qquad (4)$$

where F^a is a linear combination of structure functions with definite crossing properties, 0_i^n are local composite operators, μ is the renormalization scale and m are quark masses.

Our discussion will be limited to leading order in $1/Q^2$ and contributions from operators with twist ($\equiv$ dimension - spin) greater than two can therefore be discarded. We assume that the operators are defined by some version of minimal subtraction in which case [12, 13] the coefficients $\hat{C}$ are analytic functions of m^2 and we can replace the coefficients of the twist two operators by functions $C_i^{a,n}\left(Q^2/\mu^2, \alpha_s(\mu^2)\right)$ to leading order in $1/Q^2$. It is convenient to **define** the operators so that $C\left(1, \alpha_s(Q^2)\right) = 1 + 0(\alpha_s)$ for quark bilinears and $C = 0 + 0(\alpha_s)$ for gluon bilinears and to **define** (anti)quark distributions by

$$q_{i,n}^{\mp}(Q^2) \equiv \int_0^1 x^n \left(q_i(x, Q^2) - \bar{q}_i(x, Q^2)\right) dx$$

$$\equiv \frac{i^n}{(p_+)^{n+1}} < p | \left[\bar{\Psi}_i(0) \gamma_+ \left(\partial_+\right)^n \Psi_i(0)\right]_{Q^2} | p >_c, \qquad (5)$$

where the $-(+)$ sign applies to the case n even (odd) and $[]_{Q^2}$ denotes a Zimmermann normal product of operators defined in some version of minimal subtraction with mass scale $\sqrt{Q^2}$, which scale is also to be used for renormalizing ordinary ultraviolet divergences, calculated in light-cone $(A_+ = A_0 + A_3 = 0)$ gauge (in other gauges ∂_+ is replaced by the covariant derivative D_+).

To zeroth order in $1/Q^2$ and leading order in α_s - but only to this order - deep inelastic scattering can be described entirely probabilistically and the structure functions are related to the (anti)quark distributions defined formally by (5) in the standard way, with

$$F_2^{em}(x, Q^2) = \sum_i Q_i^2 \left(q_i(x, Q^2) + \bar{q}_i(x, Q^2) \right) + 0(\alpha_s) + 0(1/Q^2) \qquad (6)$$

etc. The leading order evolution equations for structure functions then follow from the renormalization group equations for the composite operators in (5). These equations, which control the Q^2 dependence of the $q_{i,n}^{\pm}(Q^2)$, deal separately with the non-singlet combinations of $(q_i + \bar{q}_i)$ which we write

$$(q + \bar{q})_\lambda = \sum_i \lambda_{ii}(q_i + \bar{q}_i) \qquad (7)$$

where $Tr\lambda = 0$, and the singlet combination

$$(q + \bar{q}) = \sum_i (q_i + \bar{q}_i). \qquad (8)$$

Denoting the moments of these combinations by $q_{\lambda,n}^+$ and q_n^+ respectively, the Q^2 dependence is given by

$$\frac{q_{\lambda,n}^+(Q^2)}{q_{\lambda,n}^+(Q_0^2)} = \frac{q_{i,n}^-(Q^2)}{q_{i,n}^-(Q_0^2)} = exp\left(-\int_{g(Q_0^2)}^{g(Q^2)} \frac{\gamma_{NS}^n(g')dg'}{\beta(g')} \right) \qquad (9)$$

for the non-singlet quantities, where γ_{NS}^n are the anomalous dimensions of the non-singlet operators and β is the usual beta function which controls the behaviour of the running coupling constant $\bar{g}$, and

$$S_n(Q^2) = T_g exp\left(-\int_{g(Q_0^2)}^{g(Q^2)} \frac{\gamma_S^n(g')dg'}{\beta(g')} \right) S_n(Q_0^2) \qquad (10)$$

where

$$S_n = \begin{pmatrix} q_n^+ \\ g_n \end{pmatrix}, \qquad (11)$$

g_n being the moments of the gluon distribution

$$g_n(Q^2) \equiv \int_0^1 x^n g(x, Q^2) dx, \tag{12}$$

γ_S^n is the matrix of singlet anomalous dimensions and T_g is a "g ordering operator", which is necessary because $[\gamma_S^n(g), \gamma_S^n(g')] \neq 0$ (for reviews with references where the first few terms in the perturbative expansions of $\gamma(g)$ and $\beta(g)$ may be found see [14]). It is important to realize that (9) and (10) are **exact** (although, since β and γ are not known exactly, they will only be used to finite order in practice); it is the standard relationship between the quark distributions and the structure functions (Eq (6)) which is approximate.

If we now **define**

$$< p| \left[\bar{\Psi}(0) \gamma_+ \Psi(\hat{z}) \right]_{Q^2} |p> \equiv \sum_n \frac{(z)^n}{n!} < p| \left[\bar{\Psi}(0) \gamma_+ (\partial_+)^n \Psi(0) \right]_{Q^2} |p>_c \tag{13}$$

it is obvious that we can recover Eq (1) from (5) with $\overset{(-)}{q}_i(x)$ replaced by $\overset{(-)}{q}_i(x, Q^2)$ and $\bar{\Psi}_i(\hat{z}) \gamma_+ \Psi_i(0)$ by $[\bar{\Psi}_i(\hat{z}) \gamma_+ \Psi_i(0)]_{Q^2}$. Likewise, if we **define**

$$< p|Tr\left(G^{+\nu}(0) G^+{}_\nu(\hat{z})\right)_{Q^2} |p>_c \equiv \sum_n \frac{(z)^n}{n!} < p|Tr\left(G^{+\nu}(0)(\partial_+)^n G^+{}_\nu(0)\right)_{Q^2} |p>_c$$
$$\tag{14}$$

where $G^{\mu\nu}$ is the covariant gluon field tensor and the trace is over colour indices, we can construct a formal expression for the glue distribution

$$xg(x, Q^2) = \frac{-1}{\pi p_+} \int \cos(p_+ zx) < p|Tr\left(G^{+\nu}(0) G^+{}_\nu(\hat{z})\right)_{Q^2} |p>_c dz. \tag{15}$$

Eq (15) and the Q^2 dependent version of (1) have the form we are seeking but they are not useful in practice as they stand since, although the quantities defined by (13) and (14) apparently have the form of light-cone correlation functions, they are purely formal objects: each term on the right-hand sides is a normal product with its own, n dependent, definition and the left-hand sides cannot be interpreted as matrix elements of products of operators. We shall see in the next section, however, that as long as we are only interested in working to zeroth order in $1/Q^2$, the quantity $[\bar{\Psi}_i(\hat{z}) \gamma_+ \Psi_i(0)]_{Q^2}$ in Eq (13) can be replaced by the quark correlation function smeared over distances of order $1/\sqrt{Q^2}$ transverse to the light-cone, and similarly for the gluonic quantity in (14) mutatis mutandis. First, we move onto the formal description of structure functions in higher orders in α_s.

In next to leading order the straightforward parton description breaks down. For example, in leading order

$$6F_2^{ep-en} = xF_3^{(\nu-\rho)n} - xF_3^{(\nu-\rho)p} \tag{16a}$$

$$= 2x(u + \bar{u} - d - \bar{d}) \tag{16b}$$

but (16a) is **not** true in next to leading order since the appropriate coefficient functions $C\left(1, \alpha_s(Q^2)\right)$ differ in order α_s. We could define (anti)quark distributions so that the parton model description held by definition for some particular structure function, as done by Altarelli, Ellis and Martinelli [16] who define non-singlet quark distributions in terms of the combination of structure functions which satisfy the, exact, Adler sum rule. We follow an alternative procedure [17] which is not tied to a specific process and has a direct interpretation, as we shall see in the next section, and continue to **define** the (anti)quark distributions and the gluon distribution by (5) and (14) with $C = 1 + 0(\alpha_s)$ for quark operators and $0 + 0(\alpha_s)$ for gluon operators (although it is in principle no more than a convention, the choice of procedure/prescription influences the results when the coefficient functions and the evolution of the distributions are treated to a fixed order in perturbation theory; it seems [18] that the prescription of Ref [17] is closer to the "optimum" scheme of Ref [19] than the prescription of Ref [16]).

The predictions for the moments of the singlet structure function are

$$
\begin{aligned}
F_{2,n}^S(Q^2) &\equiv \int_0^1 x^{n-2} F_2(x, Q^2) dx \\
&= C_n^S\left(\alpha_s(Q^2)\right)\left(q_{n-1}(Q^2) + \bar{q}_{n-1}(Q^2)\right) \\
&\quad + \alpha_s(Q^2) C_n^g\left(\alpha_s(Q^2)\right) g_{n-1}(Q^2) + 0(1/Q^2)
\end{aligned}
\tag{17}
$$

with $n \geq 2$ and even. The corresponding coefficients C_n are structure function dependent in the non-singlet case making a general discussion, which may be found in [14], notationally complex. For definiteness we consider a specific non-singlet given by

$$F_2^{\lambda,+}(x, Q^2) = x\left(q(x, Q^2) + \bar{q}(x, Q^2)\right)_\lambda + 0(\alpha_s) + 0(1/Q^2)$$

to leading order. The predictions for the moments are

$$
\begin{aligned}
F_{2,n}^{\lambda,+}(Q^2) &\equiv \int_0^1 x^{n-2} F_2^{\lambda,+}(x, Q^2) dx \\
&= C_n^2\left(\alpha_s(Q^2)\right)\left(q_{n-1}(Q^2) + \bar{q}_{n-1}(Q^2)\right)_\lambda + 0(1/Q^2)
\end{aligned}
\tag{18}
$$

with $n \geq 2$ and even.

Equations (17) and (18) can be written as convolutions of the form

$$F_2^S (x, Q^2) = \int_x^1 C^S \left(x/y, \alpha_s(Q^2) \right) \left(q(y, Q^2) + \bar{q}(y, Q^2) \right) dy$$

$$+ \alpha_s(Q^2) \int_x^1 C^g \left(x/y, \alpha_s(Q^2) \right) g(y, Q^2) dy + 0(1/Q^2) \qquad (19)$$

$$F_2^{\lambda,+} (x, Q^2) = \int_x^1 C^2 \left(x/y, \alpha_s(Q^2) \right) \left(q(y, Q^2) + \bar{q}(y, Q^2) \right)_\lambda dy + 0(1/Q^2) \qquad (20)$$

where $C^S(x/y, 0) = C^2(x/y, 0) = \delta(1 - y/x)$. Likewise Eqs (9) and (10) are equivalent to evolution equations of the Kogut-Susskind [20] Lipatov [21] Altarelli-Parisi [22] form

$$\frac{\partial \left(q(x, Q^2) \pm \bar{q}(x, Q^2) \right)_\lambda}{\partial \ln Q^2} = \alpha_s(Q^2) \int_y^1 f_{NS} \left(x/y, \alpha_s(Q^2) \right) \left(q(y, Q^2) \pm \bar{q}(y, Q^2) \right)_\lambda dy$$

$$(21)$$

$$\frac{\partial \left(q(x, Q^2) + \bar{q}(x, Q^2) \right)}{\partial \ln Q^2} = \alpha_s(Q^2) \int_y^1 f^S \left(x/y, \alpha_s(Q^2) \right) \left(q(y, Q^2) + \bar{q}(y, Q^2) \right) dy$$

$$+ \alpha_s(Q^2) \int_y^1 f^g \left(z/y, \alpha_s(Q^2) \right) g(y, Q^2) dy$$

$$(22)$$

with a similar equation relating the evolution of $g(y, Q^2)$ to $q + \bar{q}$ and g, and the fs are the familiar splitting functions [22] to leading order in α_s. The flavour singlet combination $q - \bar{q}$ also satisfies (21).

So far we have only recast the OPE predictions for structure functions in a suggestive parton-like form which relates their behaviour to the formal light-cone correlation functions defined by (13) and (14) to all orders in α_s. In order to make this a useful exercise, we must now uncover the meaning of the formal Q^2 dependent light-cone correlation function in interacting field theory.

4) DIAGRAMS

One way to analyze deep inelastic scattering is to separate the relevant Feynman diagrams into a calculable part controlled by short distances - corresponding to the coefficient function $C(x/y, \alpha_s)$ above - and an intractable part, controlled by long distances, corresponding to the operator matrix elements. To leading order in $1/Q^2$, this separation is equivalent to factorizing off "mass singular terms" that diverge like $ln(m^2)$ for $m \to 0$, which signal sensitivity to long distances*. In this section we will use the factorization procedure [23] to recover equations (19) and (20), and the corresponding equations for the other distributions and structure functions, and show that to zeroth order in $1/Q^2$ the formal light-cone correlation functions in (13) and (14) can be replaced by smeared quark and gluon correlation functions. Since, provided we do not make a mistake, the results must take the same form as the formal results in section 3, we shall suppress spinor, colour and flavour indices and concentrate on the general structure.

Our approach follows that of Ellis et al. [23] and subsequent authors (e.g. [17]). We start from the observation [24] that, apart from infrared divergences, singularities in scattering amplitudes can only be generated by configurations in which virtual processes may become real (the most general conditions in which singularities occur are given in [25]). We shall ignore infrared divergences here on the (insufficient) grounds that they must cancel as the structure functions are infrared finite in QCD; this simple-minded approach produces the right answer, although it requires a detailed justification which is not at all simple (see [26] for

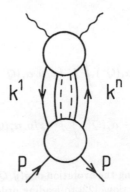

Fig 2 Diagrams which generate logs of m if $p^2 \to 0$ and all the momenta k_μ^i are proportional to p_μ.

---

* In general, sensitivity to long distances is signalled by logs of m^2 which are generated by configurations in which virtual particles get close to mass-shell and can propagate over long distances. In zeroth order in $1/Q^2$, powers of $ln(m^2/Q^2)$ occur. In next order, we encounter $m^2 \left(ln(m^2/Q^2)\right)^p /Q^2$ which is non-analytic, but not singular as $m \to 0$; separation of the configurations which generate such terms requires the full machinery of the OPE and, in contrast to the zeroth order case, it is obvious that the techniques are only applicable to deep inelastic scattering.

a discussion of the status of proofs of factorization in deep inelastic scattering and other processes). In the case of the forward virtual Compton amplitude, virtual processes can only become real if $p^2 \to 0$ and internal masses $m \to 0$ and the "mass singular" configurations (or more generally configurations that generate non-analytic behaviour) have the form shown in Fig 2 and consist of intermediate massless particles all of whose four-momenta k_μ^i are proportional to p_μ.

Dimensional analysis show that in theories without vector particles, only two particle states generate actual singularities (as opposed to non-analyticity in m) to zeroth order in $1/Q^2$; this is also true in gauge theories if light-cone gauge, or some other "physical" gauge, is employed. We therefore separate these terms explicitly as shown in Fig 3, where the bubbles marked C and those marked with a cross represent the imaginary part of the sum of all two particle irreducible diagrams*.

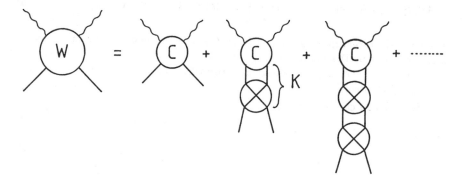

Fig 3 Expression for the imaginary part of the forward virtual Compton amplitude (W) in terms of two particle irreducible diagrams.

---

* It is at this point that we part company with the OPE which (to order $1/Q^2$) deals with terms such as $(m^2/Q^2)ln(Q^2/m^2)$, allowing the total contribution of order m^2/Q^2 to be written in the form $(m^2/Q^2)F(Q^2/\mu^2)G(m^2/\mu^2)$. This requires the separation of configurations in which three and four particle intermediate states become (almost) real. These states cannot be isolated in a clean "non-overlapping" way and the full machinery of Zimmermann's forest formula and the OPE is required: for a recent simplified discussion see Ref 13 (note that in Eq (3.4) of this paper the factor $1 - t_\epsilon$ is redundant as the factor on which it operates is finite for $\epsilon \to 0$; it would not have arisen if $\tilde{U}$ in (3.1) had been defined so that $\gamma = \gamma(m)$ was excluded).

We write the equation depicted in Fig 3 as

$$W = C\frac{1}{I - K}$$

$$= C\frac{1}{I - (I - P_\Lambda)K}\left(I + P_\Lambda K\frac{1}{I - K}\right)$$

(23)

where, as is easily verified, the second line is equivalent to the first for any operator/matrix P_Λ. Matrix multiplication in (23) is defined so that the amplitude in Fig 4 is written

$$AB = \int A(..,k)B(k,..)\frac{d^4k}{(2\pi)^4}.$$

(24)

We define P_Λ so that

$$AP_\Lambda B = \int A(..,k)\Big|_{\substack{k=xp_+ \\ k_-=\vec{k}_T=0}} dx \int B(k,..)\delta(x - k_+/p_+)\Theta(\Lambda^2 - \vec{k}_T^2)\frac{d^4k}{(2\pi)^4}$$

(25)

where p is the target momentum $p = (p_0, 0, 0, p_3)$ (a cut-off on transverse momenta was introduced by Drell and Yan [27] in the context of deep inelastic processes and was later used especially by Lepage and Brodsky [28] in their classic analysis of exclusive processes).

Equation 23 can be written

$$W(x, Q^2) = \int\limits_x^1 H(x/y, Q^2, \Lambda^2, \mu^2, m^2)M(y, \Lambda^2, m^2, \mu^2)dy + 0(p^2)$$

(26)

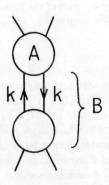

Fig 4 The quantities A and B in equations 24 and 25.

where H is generated by the term $C\left[I - (I - p_\wedge)K\right]^{-1}$, the $0(p^2)$ piece arises from setting $p_- = 0$ in the disconnected part generated by the I in the bracket in the second line of (23), and

$$M = \int T(k,p)\delta(y - k_+ /p_+)\Theta(\Lambda^2 - \vec{k}_T^2)\frac{d^4 k}{(2\pi)^4} \tag{27}$$

where $T = I + K(I - K)^{-1} = (I - K)^{-1}$ is the complete two-to-two amplitude. Provided a mass independent renormalization scheme is used (e.g. some form of minimal subtraction) the function H is free of mass singularities (it is not, however, analytic in m and contains intractable long distance contributions of order $1/Q^2$). Setting $\Lambda^2 = \mu^2 = Q^2$, Eq 26 therefore gives

$$W(x,Q^2) = \int_x^1 H\left(x/y, \alpha_s(Q^2)\right) M(y,Q^2,m^2)dy + 0(1/Q^2) \tag{28}$$

which is equivalent to Eqs (19) and (20) to leading order in $1/Q^2$, with M playing the role of q_i for a process involving a quark of flavour i. Writing T, which is the imaginary part of a forward amplitude, in terms of a matrix element of quark fields and reinstating spinor and flavour labels in conformity with the formal results in section 3, Eq (27) becomes

$$\begin{aligned} M_i(x,Q^2) &= \frac{1}{(2\pi)^2}\int e^{-ip_+ zz}dz \int \Theta(Q^2 - \vec{k}_T^2)d^2 k_T \int e^{i\vec{k}_T \cdot \vec{z}_T} \\ &\quad < p|\bar{\Psi}_i(z)\gamma_+ \Psi_i(0)|_{z_+ = 0}|p>_{c,Q^2} d^2 z_T \\ &= q_i(x,Q^2) + 0(1/Q^2) \end{aligned} \tag{29}$$

where the subscript Q^2 on the matrix element is a reminder that $\mu^2 = Q^2$.

Eq (29), and the analogous results for $\bar{q}_i(x,Q^2) = -q_i(-x,Q^2)$ and for

$$\begin{aligned} xg(x,Q^2) &= -\frac{1}{2p_+ \pi^2}\int cos(p_+ zx)dz \int \Theta(Q^2 - \vec{k}_T^2)d^2 k_T \int e^{i\vec{k}_T \cdot \vec{z}_T} \\ &\quad < p|Tr\left(G^{+\nu}(z)G^+{}_\nu(0)\right)|_{z_+ = 0}|p>_{c,Q^2} d^2 z_T + 0(1/Q^2) \end{aligned} \tag{30}$$

are the main results of this paper. We observe that -

1) The smearing function

$$J(z_T,Q^2) \equiv \frac{1}{2\pi}\int e^{i\vec{k}_T \cdot \vec{z}_T}\Theta(Q^2 - \vec{k}_T^2)d^2 k_T \tag{31}$$

that enters (29) and (30), which satisfies

$$\int J d^2 z_T = 1,$$

vanishes rapidly for $\vec{z}_T^2 > 1/Q^2$ (it tends to $\delta^2(\vec{z}_T)$ as $Q^2 \to \infty$). The quark and gluon distributions are therefore Fourier transforms along the light-cone of correlation functions smeared over distances of order $1/\sqrt{Q^2}$ transverse to the light-cone according to Eqs (29) and (30), as advertised earlier.

2) The evolution equations for the distributions can be derived from (29) and (30). Returning to (27), we have

$$\frac{\partial M(y,Q^2,m^2,Q^2)}{\partial lnQ^2} = \frac{\partial M(y,\Lambda^2,m^2,Q^2)}{\partial ln\Lambda^2}|_{\Lambda^2=Q^2} + \frac{\partial M(y,Q^2,m^2,\mu^2)}{\partial ln\mu^2}|_{\mu^2=Q^2}$$
(32).

The second term is given by the standard renormalization group equation for T and is equal to $\gamma_q(\alpha_s)M[\gamma_g(\alpha_s)M]$ in the case of quarks [gluons] where $\gamma_q[\gamma_g]$ is the anomalous dimension of the quark [gluon] field. The first term is given by

$$\frac{\partial M}{\partial ln\Lambda^2}|_{\Lambda^2=Q^2} = \int \delta(x - k_+/p_+)\delta(\vec{k}_T^2 - Q^2)T_c\frac{d^4k}{(2\pi)^4}$$
(33)

where $T_c = K(I - K)^{-1}$ is the connected two-to-two amplitude. Setting*

$$T_c = K\frac{1}{I - (I - P_{\bar{\lambda}}K)}[I + P_{\bar{\lambda}}T_c]$$

in (33), using the fact that $K[I - (I - P_\lambda)K]^{-1}$ is free of mass singularities, and setting $\bar{\Lambda}^2 = Q^2$ we obtain the evolution equation

$$\frac{\partial M(y,Q^2)}{\partial lnQ^2} = \alpha_s(Q^2)\int_y^1 f\left(y/z,\alpha_s(Q^2)\right)M(z,Q^2)dz + 0(1/Q^2),$$
(34)

where the second term in (32) contributes a piece $\gamma\delta(z - y)$ to f, which is equivalent to (21) and (22).

3) Eqs (29) and (30) are not manifestly gauge invariant. To restore manifest invariance, we would have to insert an additional factor

$$Pexp\left(ig\int_0^z A_\mu(z')dz'_\mu\right).$$

---

* The form given here appears *ab initio* in Eq (23) in the literature and it has generally not been observed that it is equal to T_c; this (together with the fact that the structure is more complicated if a projection operator P_ϵ with a dimensional cut-off is used in place of P_λ) is perhaps why the fact that M in (26) is a smeared correlation function has not been pointed out earlier.

The integral in this factor is equal to $-\int_0^z \vec{A}_T . d\vec{z}_T$ because $z_+ = 0$ and $A_+ = 0$.

Since z_T is of order $1/\sqrt{Q^2}$, the additional factor only changes the result by terms of order $1/Q^2$ (using the fact that the distributions are even functions of g) and can therefore be ignored.

4) The coefficients in the convolution equations (19), (20), (28) and the evolution equations (21), (22), (34) are scheme dependent beyond leading order - they depend on the (mass independent) renormalization scheme ($MS, \overline{MS}$...) and also on the precise form of the cut-off in the projector P_Λ, which we defined by (25) where, for example, the Θ function could be replaced by a smooth cut-off function or a covariant cut-off could have been introduced.

5) DISCUSSION

In the previous section we

a) rederived the all orders in α_s QCD corrected parton model, which expresses structure functions as convolutions of distribution functions and "hard scattering amplitudes" (which however do not have a simple interpretation as they include projection operators), and

b) showed that the quark and gluon distributions are Fourier transforms along the light-cone of correlation functions smeared over distances of order $1/\sqrt{Q^2}$ transverse to the light-cone.

We now consider the implications of the latter result for model calculations of structure functions and for studies of the distances that control deep inelastic scattering.

It is important to realize that $q(x, Q^2)$ and $g(x, Q^2)$, which we **defined** in terms of the matrix elements of twist two operators, are only equal to the transforms of smeared correlation functions up to order $1/Q^2$ (as indicated in Eq (29)). The basic difference is that twist two operators are rendered finite by subtractions whereas the smeared correlation functions are rendered finite simply by setting $\vec{k}_T^2 \leq Q^2$ in the integral over the momenta of the two external $q\bar{q}$ or gg legs. It is apparent in the fact that whereas $\int (q - \bar{q}) dx$ is given by the matrix element of $\bar{\Psi}(0)\gamma_+ \Psi(0)$, which is a special case and requires no subtractions, and is equal to 3 with no corrections, the integral of the corresponding smeared quantity is not exactly equal to 3. This difference has what appears at first sight to be a most unfortunate consequence. If we knew $q(x, \mu^2)$ and $g(x, \mu^2)$ we could use the **exact** evolution equations (21) and (22) to predict $q(x, Q^2)$ and $g(x, Q^2)$ up to corrections of order $\left(\alpha_s(Q^2) - \alpha_s(\mu^2)\right)^{n+1}$, assuming all perturbative calculations have been carried out to n^{th} order, from which we could then calculate the structure functions $F(x, Q^2)$ up to additional corrections of order $\left(\alpha_s(Q^2)\right)^{n+1}$ and $1/Q^2$. If on the other hand we started from the smeared correlation functions at a scale μ^2, there would be further corrections of order $1/\mu^2$ due to the fact that the evolution equation for $M(\mu^2)$ has corrections of this order. Unfortunately, although it is easy to imagine calculating or modelling $M(\mu^2)$, it is very hard to imagine calculating the twist two matrix elements, which are (almost) all ultra-violet divergent and require scheme dependent renormalizations. Even if we had available the exact solution of QCD, we could only extract the twist-two matrix elements by subtracting the $0\left(1/(Q^2)^p\right), p \geq 1$, part of the moments of the structure functions and dividing out the coefficient functions.

Fortunately the situation is actually much better than it might appear. The first element of good news is that in models with an intrinsic cut-off, which itself produces a smearing, Eq (1) can be used directly and it is not necessary to perform the smearing in Eq (29), which would only change the order $1/\mu^2$ error. Indeed, in models that are constructed so that conserved quantities such as $\bar{\Psi}(0)\gamma_+\Psi(0)$ have the correct matrix elements, Eq (1) correctly predicts the corresponding structure function moments at all Q^2, up to corrections of order $\left(\alpha_s(Q^2)\right)^{n+1}$ and $1/Q^2$: smearing would gratuitously introduce additional errors of order $1/\mu^2$ and is therefore best avoided. Second, it can be argued that when Eq (1) is used to construct $q(x, \mu^2)$ in such models, the $0(1/\mu^2)$ errors are likely to be numerically very small.

In quark, bag, lattice or other models endowed with an implicit or explicit ultraviolet cut-off Λ Eq (1) is well defined and so are the twist two matrix elements, which are given exactly by the moments of the expressions in Eq (1) when ultraviolet divergences are eliminated. It is plausible that, as argued by Jaffe and Ross [29], such models produce results for twist two matrix elements at an unknown scale μ, which might be several hundred MeV in quark or bag models. The scale μ at which the models are deemed to apply must obviously be much less than the implicit or explicit cut-off Λ but it follows from the renormalization group that μ/Λ must remain fixed if Λ is varied i.e. we have $\mu = \epsilon\Lambda$ with $\epsilon << 1$. In addition to errors reflecting imperfections in the model, all predictions will in general contain errors of order m^2/Λ^2, where m is some typical hadronic scale e.g. $m^2 = <\vec{p}_T{}^2>$. In the case of twist two matrix elements these are the $0(1/\mu^2)$ errors encountered above which, we now see, are in fact likely to be of order $m^2/\Lambda^2 = \epsilon^2 m^2/\mu^2$ and therefore unimportant compared to other errors introduced by the model.

The prescription for modelling structure functions which emerges is therefore -
1) Calculate q and $\bar{q}$ using Eq (2) as advocated by Signal and Thomas [4] who observed that Eq (2) leads to distributions which automatically have the right support and satisfy $\bar{q} > 0$, in contrast to results obtained in earlier calculations *. The model calculation should if possible give the correct results for

---

* Earlier bag calculations [30] led to structure functions with the (wrong) support $0 < x < \infty$ because the approximations used destroy translational invariance. A prescription, inspired by results obtained in a translationally invariant model in one space dimension [31], is generally used to map the region $[0, \infty]$ to $[0, 1]$ - see [32]. The original calculations of Jaffe [30] also led to a negative 'sea', a disease attributed to the neglect of bubble graphs which however generate a divergent sea in the rigid cavity approximation [33].

conserved quantities.

2) Following Jaffe and Ross [29], interpret q and $\bar{q}$ as giving twist two matrix elements at some (unknown but presumably relatively small) scale μ, evolve q and $\bar{q}$ to higher Q^2 and then convolute them with the appropriate coefficient functions to give structure functions. There will be a) intrinsic errors due to imperfections in the model and the uncertainty in choosing μ, b) order $\epsilon^2 m^2/\mu^2$ errors due to the necessity of using a cut-off in calculating twist two matrix elements, c) errors of order m^2/Q^2 due to higher twist contributions, and d) errors of order $\left(\alpha_s(Q^2) - \alpha_s(\mu^2)\right)^{n+1}$ and $\left(\alpha_s(Q^2)\right)^{n+1}$ due to truncation of the perturbative calculation of the evolution kernels and the coefficient functions.

Finally we consider the interpretation of the smeared correlation functions and the question of the distances that play an important role in deep inelastic scattering. Defining $\tilde{z}_\mu = (z, \vec{z}_T, -z)$, the quark correlation function satisfies

$$< p|\Psi_+^\dagger(\tilde{z})\Psi_+(0)|p>_c = - <p|\Psi_+(0)\Psi_+^\dagger(\tilde{z})|p>_c$$

for $|\vec{z}_T| \neq 0$, because $\tilde{z}$ is space-like, and also for $\vec{z}_T = 0$, because the anticommutator of Ψ_+ and $\Psi_+^\dagger$ is a C number for $z_+ = \vec{z}_T = 0$ as discussed in section 2. Consequently the discussion in section 2 requires little modification when (29) replaces (1), the only change in Eq (2) being that the sum should only include states n which satisfy $(\vec{p}_T^n)^2 \leq Q^2$ (in a frame in which $\vec{p}_T = 0$ for the target). In particular, it is clear that the part of the smeared correlation function which contributes to $q(x, Q^2) \left[\bar{q}(x, Q^2)\right]$ for $x \geq 0$ is the amplitude to destroy a quark [antiquark] at the origin and reinstate it at $\tilde{z}$ leaving the target intact, smeared over $\vec{z}_T^2 \lesssim 1/Q^2$. This quantity gives a measure of the extension of the target as seen by a relativistic projectile.

Given a set of [anti] quark distributions, it is a simple matter to invert the Fourier transform in (29) and construct the smeared correlation function. Defining

$$C(z) \equiv \frac{1}{2\pi} \int J(z_T, Q^2) <p|\Psi_+^\dagger(\tilde{z})\Psi_+(0)|p> d^2 z_T,$$

where J is given by (31), it is convenient to consider separately the quantities

$$C_\pm(z) \equiv \frac{C(z) \pm C(-z)}{2}$$

which are given by cosine and sine transforms of $q \pm \bar{q}$ respectively. Figs 5 and 6 show the results obtained [5] for a nucleon at rest with a (rather arbitrary) "valence" distribution $q_V = q - \bar{q} \sim (1 + 4x)(1 - x)^3/\sqrt{x}$ and a "sea" distribution $(q + \bar{q})_S \sim (1 - x)^5/x$ (note that $C_+(0)$ is equal to the normalization of the valence distribution). The large z behaviour is highly dependent on the behaviour of $q \mp \bar{q}$ for $x \to 0$, the range Δz of $C_\pm(z)$ being related to the extent (Δx) of the spike for $x \to 0$ by $\Delta z \sim 1/(M_N \Delta x)$, as anticipated in section 2, with $\Delta x = 0$ for a $1/x$

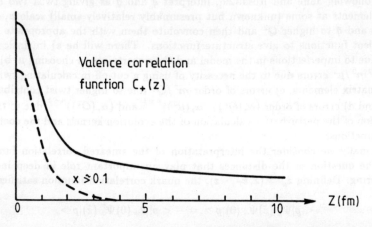

Fig 5 The smeared correlation function C_+ obtained by inverting (29) with a typical shape for $q - \bar{q}$. The dashed curve was obtained by eliminating the spike for $x \to 0$ while leaving the $x \gtrsim 0.1$ region essentially unchanged.

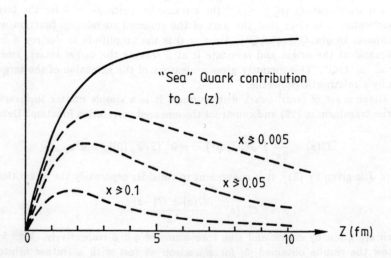

Fig 6 The "sea" quark contribution to C_- obtained by inverting (29) with a typical shape for the $q + \bar{q}$ sea. The dashed curves were obtained by eliminating the spike for $x \to 0$ while leaving the $x \gtrsim 0.005$, $x \gtrsim 0.05$, or $x \gtrsim 0.1$ regions essentially unchanged.

distribution. Physically the long range correlation can be understood by returning to deep inelastic scattering and considering the diagram in Fig 7 in Rayleigh-Schrödinger perturbation theory. Assuming that the invariant mass M of the intermediate $q\bar{q}$ state stays finite, the energy difference $\Delta E = q_0 - \sqrt{M^2 + \vec{q}^2} \approx x M_N$ tends to zero as $x \to 0$ in the Bjorken limit, allowing the intermediate state to "live" for an infinite time and therefore generate an infinite correlation length. The results of eliminating the small x contributions (by multiplying by $\left(x/(x^4 + x_c^4)^{1/4}\right)^{1/2}$ which suppresses the region $x \lesssim x_c$ leaving the region $x \gtrsim x_c$ essentially unchanged) are shown in Figs 5 and 6. We see that for $x \gtrsim 0.1$ the valence distribution is controlled by the correlation function for $z \lesssim 1 fm$ whereas very much greater distances play a role in the sea distribution.

It follows from this analysis that whereas it may make sense to discuss the gross features for the EMC effect in terms of modifications of the properties of single nucleons in the nucleus for $x \gtrsim 0.1$, a single nucleon treatment must fail for $x \lesssim 0.1$ where the relevant light-cone correlation lengths are larger than the typical internucleon separation. If the valence and sea contributions to the EMC effect had been separated experimentally, we could construct the difference between nuclear (A) and nucleon (N) correlation functions. As an illustration, we show $C_+^A - C_+^N$ corresponding to the very crude model function $q_V^A - q_V^N \sim (0.2 - x)(1 - x)^3$ which has the correct feature that it integrates to zero (it is clearly wrong for $x \gtrsim 0.8$, where the actual function - which extends beyond one - changes sign, but the distributions are very small at large x). The result, which is shown in Fig 8 with and without the small x region being cut out, confirms that for $x \gtrsim 0.1$ the EMC effect is generated by an increase in the light-cone correlation function for $z \sim 1 fm$. The magnitude of this increase is, however, only about 1%.

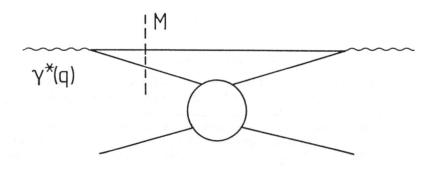

Fig 7   A contribution to the forward virtual Compton amplitude.

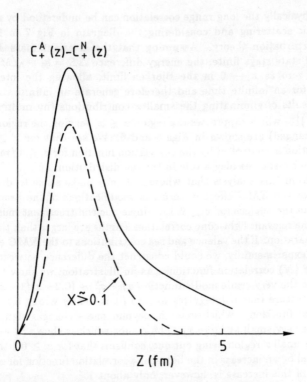

Fig 8 The difference between the smeared "valence" correlation functions for nuclei and nucleons corresponding to the model function for $q_V^A - q_V^N$ in the text. The dashed curve was obtained by suppressing the region $x \lesssim 0.1$. The maximum of $C_+^A - C_+^N$ here is only about 1% of $C_+(0)$ in Fig 5.

Acknowledgements

I am very grateful to Tony Thomas for discussions about model calculations of structure functions and about the *EMC* effect during a most enjoyable visit to Adelaide and his return visit to Oxford. These discussions encouraged me to write down the explicit connection between structure functions and smeared correlation functions and develop the treatment presented here to all orders in α_s. I am also grateful to Gabriel Karl, Boris Castel and Pat O'Donnell for their hospitality at the CAP-NSERC Summer Institute and to Bob Jaffe for comments on the typescript.

References

1) Brandt, R.A., Phys. Rev. Lett. **22**, 1149 (1969) and **23**, 1260 (1969)
   Ioffe, B.L., Zh. Eksp. i Theor. Fiz Pisma **9**, 163 (1969) [JETP Letters **9**, 97 (1969)]; Phys. Lett. **B 30**, 123 (1969)
   Brown, L.S. in "Proc. Boulder Conference on High Energy Physics", ed. K.T. Mahanthappa, W.D. Walker and W.E. Brittin, Colorado Associated University Press (1970)

2) Jackiw, R, Van Royen, R. and West, G.B., Phys. Rev. **D2**, 2473 (1970)
   Fritzsch, H. and Gell-Mann, M., in "Coral Gables Conference on Fundamental Interactions at High Energy", ed. M.D. Cin, G.J. Iverson and A. Perlmutter, Gordan and Breach (New York), (1971)
   Cornwall, J.M. and Jackiw, R., Phys. Rev. **D4**, 367 (1971)

3) Jackiw, R. and Preparata, G., Phys. Rev. Lett. **22**, 975 (1969) and **22**, 1162(E) (1969), and Phys. Rev. **185**, 1748 (1969)
   Adler, S.L. and Wu-Ki Tung, Phys. Rev. Lett. **22**, 978 (1969) and Phys. Rev. **D1**, 2846 (1970)

4) Signal, A.I. and Thomas, A.W., Phys. Lett. B (in press)
   Thomas, A.W., to be published by AIP Press in Proc. Argonne meeting on Nuclear Chromodynamics, May 1988

5) Llewellyn Smith, C.H., Nucl. Phys. **A434**, 35c (1985)

6) Collins, J.C. and Soper, D.E., Nucl. Phys. **B194**, 445 (1982)

7) Jaffe, R.L., Nucl. Phys. **B229**, 205 (1983)

8) Zimmermann, W., in "Lectures in Elementary Particles and Quantum Field Theory", ed. S. Deser, M. Grisaru and H. Pendleton, MIT Press (Cambridge Mass.) (1971)

9) Wilson, K., Cornell Report LNS-64-15, (1965) (unpublished) and Phys. Rev. **179**, 1499 (1969)

10) Brandt, R.A., Ann. Phys. (NY) **44**, 221 (1967)

11) Christ, N., Hasslacher, B. and Mueller, A.H., Phys. Rev. **D6**, 156 (1972)
    Cornwall, J. and Norton, R., Phys. Rev. **177**, 2584 (1969)

12) Tkachov, F.V., Phys. Lett. **124B**, 212 (1983)
    Gorishny, S.G., Larin, S.A. and Tkachov, F.V., Phys. Lett. **124B**, 217 (1983)

13) Llewellyn Smith, C.H. and de Vries, J.P., Nucl. Phys. **B296**, 991 (1988)

14) Furmanski, W. and Petronzio, R., Z. Phys. **C11**, 293 (1982)
    Buras, A.J., Rev. Mod. Phys. **52**, 199 (1980)

15) Llewellyn Smith, C.H., Nucl. Phys. **B17**, 277 (1970)

16) Altarelli, G., Ellis, R.K. and Martinelli, G., Nucl. Phys. **B143**, 521 (1978)

17) Curci, G., Furmanski, W., and Petronzio, R., Nucl. Phys. **B175**, 27 (1980)

18) Chýla, J., Dubna preprint E2-88-293 (1988)

19) Stevenson, P.M. and Politzer, H.D., Nucl. Phys. **B175**, 758 (1986)

20) Kogut, J. and Susskind, L., Phys. Rev. **D9**, 697 and 3391 (1964)

21) Lipatov, L.N., Sov. J. Nucl. Phys. **20**, 94 (1975)

160

22) Altarelli, G. and Parisi, G., Nucl. Phys. **B126**, 298 (1977)
23) Amati, D., Petronzio, R. and Veneziano, G., Nucl. Phys. **B140**, 54 (1978); **B146**, 29 (1978)
   Libby, S. and Sterman, G., Phys. Rev. **D18**, 3252 (1978)
   Mueller, A., Phys. Rev. **D18**, 3705 (1978)
   Ellis, R., Georgi, H., Machacek, M., Politzer, H.D. and Ross, G.G., Nucl. Phys. **B152**, 285 (1979)
   Efremov, A. and Radyushkin, A., Theor. Math. Phys. **44**, 664, 774 (1981)
24) Coleman, S. and Norton, R.E., Nuovo Cimento **XXXVIII**, 438 (1985)
25) Sterman, G., Phys. Rev. **D17**, 2773 and 2789 (1978)
26) Collins, J.C. and Soper, D.E., Ann. Rev. Nucl. Sci. **37**, 383 (1987)
27) Drell, S.D. and Yan, T-M., Phys. Rev. Lett. **24**, 181 (1970) and Ann. Phys. (NY) **66**, 578 (1971)
28) Lepage, G.P. and Brodsky, S.J., Phys. Rev. **D22**, 2157 (1980)
29) Jaffe, R.L. and Ross, G.G., Phys. Lett. **93B**, 313 (1980)
30) Jaffe, R.L., Phys. Rev. **D11**, 1953 (1975)
31) Krapchev, V., Phys. Rev. **D13**, 329 (1976)
32) Jaffe, R.L., Ann. Phys. (NY) **132**, 32 (1981)
33) Bell, J.S., Davis, A.C. and Rafelski, J., Phys. Lett. **78B**, 67 (1978)

THE EMC EFFECT AND THE ORIGIN OF MASS

J.V. Noble[*]

Institute of Nuclear and Particle Physics
Department of Physics, University of Virginia
Charlottesville, Virginia 22901

ABSTRACT

This paper studies the celebrated EMC effect. We consider the relation between deep-inelastic structure functions and a quark-level description of the nucleon. We show the quark-knockout picture describes both valence and "sea" quark distributions. This approach avoids the conceptual diffi-culties of the infinite momentum frame, especially the loss of binding effects.

We then study how the structure function of a bound nucleon is related to that measured for an isolated nucleon. The usual method of expressing the scaling variable x as a ratio of Lorentz invariants does not apply to bound (hence, off-shell) particles. A prescription is given based on the elementary Doppler effect, assuming the velocity distribution of the nucleus is known. This technique makes clear that the EMC effect is not a simple nuclear binding effect. Since both the Doppler method and the usual method lead to the convolu-tion formula of the light-cone formalism, the latter reveals nothing about binding effects.

Finally we show that the EMC effect can be explained as a softening of the quark momentum distribution of bound nuc-leons, analogous to the softening of the elastic form factors of bound nucleons. This suggests that the underlying mechanism of confinement is controllable by the experimental-ist, through variation of nuclear density. This will permit the elucidation of the origin of the hadron masses.

CONTENTS

1. Introduction
2. Quark models of the structure functions
3. Fermi motion as a Doppler effect
4. What is the EMC Effect?
5. Relation to the origin of mass

[*]Supported in part by the US National Science Foundation

1. Introduction

Because we have, at present, no fundamental theory of nuclear structure (nor, for that matter, a theory of *nucleon* structure) in terms of
QCD degrees of freedom, attempts to relate deep-inelastic scattering
from nuclei to deep-inelastic scattering from isolated nucleons
possess conjectural elements. Intuitively we expect a weakly bound
system hardly differs from an equivalent number of free particles. Had
it turned out that the nuclear cross-sections were merely the atomic
weight, A, times the isolated-nucleon cross-sections, we should have
considered intuition vindicated and gone on to more rewarding
investigations. However, as first reported by the European Muon
Collaboration[1] (EMC) and subsequently verified by groups at SLAC[2] and
at CERN[3], the ratio of the structure function, $F_2(x,Q^2)$, for a nucleus
to that of Z protons and N neutrons differs markedly from unity
(Fig. 1), the disagreement becoming more marked — as might be
expected — with increasing A. This discrepancy is called the EMC
effect.

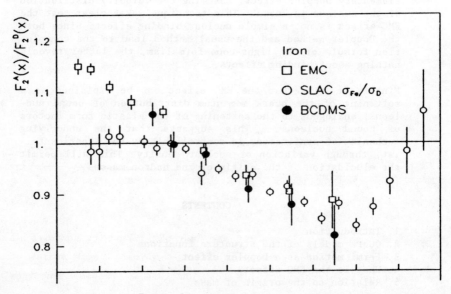

Fig. 1 *Ratio of $F_2^{Fe}(x)/F_2^d(x)$ as measured by EMC, et al.*

We remind the reader what precisely is measured in deep-inelastic scattering: Recall that the inclusive inelastic cross section can, in one-photon exchange approximation, be written

$$\frac{d^2\sigma}{d\Omega dE'} \propto w_{\mu\nu} W^{\mu\nu} \tag{1}$$

where $W^{\mu\nu}$ is a symmetric tensor representing all the hadronic information:

$$W^{\mu\nu} = \sum_X \langle X|J^\mu|i\rangle \langle X|J^\nu|i\rangle^* \delta(E_i + \omega - E_X) , \tag{2}$$

where $|i\rangle$ is the hadronic target state, $\langle X|$ is the (unobserved) hadronic final state, and ω is the energy brought in by the virtual photon.

On general symmetry grounds $W^{\mu\nu}$ can be decomposed into two conventional tensors:

$$W^{\mu\nu} = \left[-g^{\mu\nu} + \frac{q^\mu q^\nu}{q^2} \right] W_1 + \left[p^\mu - \frac{p \cdot q}{q^2} q^\mu \right] \left[p^\nu - \frac{p \cdot q}{q^2} q^\nu \right] \frac{W_2}{M^2} , \tag{3}$$

where

$$q = \begin{pmatrix} \omega \\ \vec{q} \end{pmatrix} \tag{4}$$

is the 4-momentum carried by the virtual photon, and p is that of the target.

In the limit as $Q^2 = -q^2 \rightarrow \infty$, if the hadronic target is composed of strictly pointlike objects, all masses and internal length scales of the target become irrelevant, and the functions $W_{1,2}$ become functions of a dimensionless ratio

$$x = \frac{Q^2}{2M\omega} \ . \tag{5}$$

In fact, we have

$$MW_1(\omega, \ Q^2) \xrightarrow[Q^2 \to \infty]{} F_1(x) \tag{6a}$$

$$\omega W_2(\omega, \ Q^2) \xrightarrow[Q^2 \to \infty]{} F_2(x) \ . \tag{6b}$$

Suppose the target is a stationary proton; then

$$W^{00} = \frac{\vec{q}^2}{Q^2}\left[-W_1 + \frac{\vec{q}^2}{Q^2}W_2\right] \tag{7}$$

From the definition Eq. 5

$$|\vec{q}|^2 = \left[\omega + Mx\right]^2 - M^2x^2 \ , \tag{8}$$

hence

$$\frac{\vec{q}^2}{Q^2} = \frac{\omega}{2Mx} + 1 \tag{9}$$

and

$$W^{00} = \left[\frac{\omega}{2Mx} + 1\right]\left[\left[\frac{\omega}{2Mx} + 1\right]W_2 - W_1\right] \tag{10}$$

From Eq. 10 it is obvious that for fixed x and $Q^2 \to \infty$, $\omega \to \infty$ also; hence if W^{00} is to remain finite, it must be true that

$$\frac{\omega}{2Mx}W_2(\omega, \ Q^2) \xrightarrow[Q^2 \to \infty]{} W_1(\omega, \ Q^2) \tag{11}$$

or

$$F_1(x) = 2xF_2(x) \tag{12}$$

Equation 12 is known as the Callan-Gross relation[4]. It is frequently stated that Eq. 11 is true if and only if the pointlike constituents of the nucleon have spin $\frac{1}{2}$, so that the empirical truth of Eq. 12 proves experimentally that they do indeed have spin $\frac{1}{2}$. To see why this is so, consider a boson current: it always has a piece proportional to $(2p + q)^{\mu}$ times the normalization of the initial and final states, $\sim (2\epsilon_i 2\epsilon_f)^{-1/2}$. Thus the boson current will contribute to W^{00} a term proportional to (assuming the initial boson at small momentum)

$$W^{00} \approx \delta(\epsilon_i + \omega - \epsilon_f)\frac{(\epsilon_i + \epsilon_f)^2}{4\epsilon_i \epsilon_f} \propto \omega .$$ (13)

That is, for point bosonic scatterers $W^{00} \rightarrow \infty$. Compare this with the current of a spin-$\frac{1}{2}$ point particle: here

$$W^{00} \approx \sum_f \delta(\epsilon_i + \omega - \epsilon_f) \rightarrow const.$$ (14)

The difference arises entirely from whether the Lagrangian is linear or quadratic in momenta, i.e. fermionic or bosonic.

2. Quark models of the structure functions

Suppose the dominant process in deep-inelastic scattering is quark "knockout". That is, a bound quark with momentum $\vec{p}$ absorbs a virtual photon, thereby gaining a large 3-momentum $\vec{q}$ and acting as though it were on-shell with energy $|\vec{p}+\vec{q}|$. Of course, the quark actually cannot be knocked out: rather it radiates $q\bar{q}$ pairs (mesons) until it and the recoiling (colored) remainder of the nucleon can recombine to make a baryon. This is illustrated in Fig. 2 below.

From Eq. 10 we see that

$$W^{00} \rightarrow \frac{F_2(x)}{2Mx}$$ (15)

so we can calculate $F_2(x)$ from W^{00} alone, *i.e.*, from what nuclear physicists call the *longitudinal structure function* of the target. Thus, for a stationary proton we have

$$F_2(x) = 2Mx \sum \delta(M + \omega - E_X) |\langle X|J^0(\vec{q})|P\rangle|^2 . \qquad (16)$$

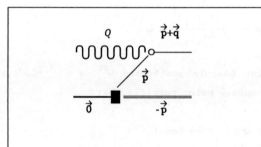

Fig. 2   *Quark-knockout as the dominant inelastic process at large Q^2*

To evaluate Eq. 16 we need the matrix elements $\langle X|J^0(\vec{q})|P\rangle$, meaning we need to know both the structure of the proton ground state $|P\rangle$ and that of the (highly excited) state $|X\rangle$, in terms of quark degrees of freedom.

The intractibility of Eq. 16 for evaluating $F_2(x)$ has led to a formalism introduced by Feynman[5], which involves working in the so-called infinite-momentum frame.   In this frame, the interactions between partons become simple, in fact can be ignored[6]; with each parton assumed to have a fraction x of the proton's total 4-momentum P. The advantages of treating the constituents as free particles is obvious, even if we know there is something wrong with neglecting an infinite confining interaction!   The disadvantage is that the structure functions then lose their relationship with the quark components of the nucleon ground-state wave function, so the connection between low and high-momentum aspects of baryon structure is lost.

The infinite momentum frame is clearly inappropriate to study the famous EMC effect, that takes place in a nucleus and that may have something to do with either nuclear binding or the modification of nucleons by the nuclear environment, since all memory of finite interactions is lost in that frame; and since we should not know how the convoluted structure functions are related to nuclear physics. Thus we shall try to understand the EMC effect in the laboratory frame.

As practice for remaining in the lab frame, let us try to evaluate $\langle X|J^0(\vec{q})|P\rangle$ using a plausible model such as a bag or hybrid model. (Jaffe[7] has previously given corresponding results from the MIT Bag model.) The momentum-space wave function of a bound (valence) quark has the form

$$\psi(\vec{p}) = \begin{pmatrix} a(\vec{p}) \\ \vec{\sigma}\cdot\hat{p}\,b(\vec{p}) \end{pmatrix} .$$ (17)

If we assume the final state $|X\rangle$ is a simple excitation of a valence quark to a state of large momentum (that is, quark "knockout") the momentum-space wave function of the final quark is

$$u(\vec{p}') = \frac{1}{\sqrt{2}} \begin{pmatrix} 1 \\ \vec{\sigma}\cdot\hat{p}' \end{pmatrix} .$$ (18)

Hence the matrix element becomes

$$\langle X|J^0(\vec{q})|P\rangle = \frac{1}{\sqrt{2}}\left[a(\vec{p}) + b(\vec{p})\hat{p}\cdot\hat{q} + ib(\vec{p})\vec{\sigma}\cdot\hat{q}\times\hat{p} \right] .$$ (19)

The contribution of Eq. 19 to W^{00} clearly has the form

$$\int d^3 p\, \varphi(\vec{p})\,\delta(M + \omega - E_X)$$ (20)

where $\varphi(\vec{p})$ is the square of Eq. 19 (summed over spins), and the energy of the intermediate state, E_X, is given approximately by

$$E_x = |\vec{p} + \vec{q}| + \sqrt{M^2 + \vec{p}^2} \,. \tag{21}$$

The contribution of Eq. 20 to W^{00} is clearly finite as $x \to 0$, hence the valence-quark contribution to $F_2(x)$ vanishes linearly with x at small x.

Contributions from the (filled) negative-energy quark states in the bag (that is, $q\bar{q}$ pair production) require $|X\rangle$ to have the form

$$|X\rangle = a^{\dagger}_{p'} b_p |P\rangle \tag{22}$$

where $a^{\dagger}$ is a positive-energy creation operator, and b is a negative-energy destruction operator. The diagrammatic representation of excited states of the above type (Eq. 22) is shown in Fig. 3 below:

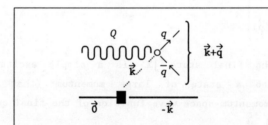

Fig. 3 $q\bar{q}$ production as a sub-dominant inelastic process at large Q^2: ... ... stands for glue exchange

The object denoted by a dashed line, carrying 3-momentum $\vec{k}$ must be a multi-gluon configuration in order that it carry no color. The process illustrated in Fig. 3 above leads to a longitudinal structure function W^{00} that diverges at small x, like x^{-1}. This follows from doing the integral (approximate structure function)

$$\int d^3k f(k^2) \int d^3p_1 \int d^3p_2 \,\delta(\omega - p_1 - p_2)\, \delta(\vec{q} + \vec{k} - \vec{p}_1 - \vec{p}_2)\, k^{-2} |N|^2 \left[q^2 - 2p_1 \cdot q \right]^{-2} \tag{23}$$

where $|N|^2$ depends only on angles, and $f(k^2)$ is the momentum distribution of the glue field. The leading term of the integral at small x is proportional to $(Mx)^{-1}$; thus its contribution to $F_2(x)$ goes to a constant at small x. (Note there is only one power of k^{-2} in Eq. 23 because the $\gamma \rightarrow q\bar{q}$ amplitude in a scalar potential carrying momentum $\vec{k}$ vanishes at $\vec{k} = 0$.) Hence the contribution of Fig. 3 to $F_2(x)$, the so-called "sea"-quark contribution[8], does not vanish at x = 0.

One may imagine the momentum distribution of the glue to fall more-or-less like the square of a nucleon elastic form factor, i.e. like k^{-8} at large k. Since the kinematics demand that k exceed

$$k_{min} = \frac{1}{2}M \left| \frac{1}{1-x} + x - 1 \right| , \qquad (24)$$

the behavior of the sea-quark contribution near x = 1 is $\approx (1-x)^8$.

As seen in Fig. 4 below,

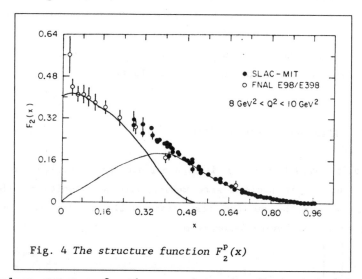

Fig. 4 *The structure function* $F_2^p(x)$

the nucleon structure function consists of a flavor-independent part that is finite at x=0 and falls rapidly with x; and a flavor-dependent contribution (the "non-singlet" part), proportional to the difference between proton and neutron structure functions, that vanishes at $x=0$ and falls more slowly with x. Thus the behavior near $x=1$ is dominated by the non-singlet part.

The above simple discussion shows that the usual attribution of a structure function into "valence" and "sea" contributions is physically reasonable as well as understandable in terms of quark wave functions of the nucleon, treated in the laboratory frame. That is, there is no reason to prefer the infinite momentum frame.

A crucial feature to bear in mind, in the above quasielastic view of deep-inelastic scattering, is that the momentum falloff of either the quark or glue momentum distributions is conditioned by the size of the target hadron. That is, the dynamical effects that lead to confinement (and to a string tension[9]) provide a momentum scale m. Since energy conservation leads to the presence of M (the nucleon mass) in Eq. 24 above, $F_2(x)$ contains the ratio m/M. It is therefore untrue to say that $F_2(x)$ is some pure mathematical function of x (except in the deep sense that the same dynamics that gives M provides the momentum cutoff m in the quark and glue wave functions).

3. <u>Fermi motion as a Doppler effect</u>

The scaling variable Eq. 5 is appropriate to a nucleon at rest. Suppose the target nucleon were moving with a velocity $\vec{v}$: then an experimenter in the nucleon rest frame will perceive the incident virtual photon as having 4-momentum

$$q' = \begin{pmatrix} \omega' \\ \vec{q}' \end{pmatrix} . \qquad (25)$$

where

$$\vec{q}' = \gamma\left[\vec{q} - \omega\vec{v}\right] \tag{26a}$$

$$\omega' = \gamma\left[\omega - \vec{v}\cdot\vec{q}\right] , \tag{26b}$$

and

$$\gamma = \left[1 - \vec{v}^2\right]^{-1/2} . \tag{27}$$

He will then measure a Doppler-shifted structure function $F_2(x', Q^2/\Lambda^2)$, where

$$x' = \frac{Q'^2}{2M\omega'} \equiv \frac{Q^2}{2M\omega'} . \tag{28}$$

Since, for a free nucleon at velocity $\vec{v}$,

$$\begin{pmatrix} p^0 \\ \vec{p} \end{pmatrix} = M\gamma \begin{pmatrix} 1 \\ \vec{v} \end{pmatrix} , \tag{29}$$

we see that Eq. 28 can be re-written in terms of Lorentz invariant quantities:

$$x' = \frac{Q^2}{2p\cdot q} . \tag{30}$$

The simplicity and appealing form of Eq. 30 have led to its (mis)application to the Fermi averaging of bound nucleons[10]. We suppose a nucleon bound in a nucleus has the momentum distribution $n(\vec{p})d\vec{p}$ and take for the Fermi-averaged structure function

$$\langle F_2(x)\rangle = \int d\vec{p}\, n(\vec{p}) F_2(x') \approx \int d\vec{p}\, n(\vec{p}) F_2\left(\frac{xM}{p^0 - \vec{p}\cdot\hat{q}}\right) \tag{31}$$

where we have derived Eq. 31 using the relation (by definition)

$$|\vec{q}|^2 = \left(\omega + Mx\right)^2 - M^2x^2 , \tag{32}$$

to write

$$|\vec{q}|/\omega = 1 + O(M/\omega) \quad ; \tag{33}$$

together with the laboratory scaling variable, Eq. 5.

As long as we are dealing with a nuclear model in which nucleons have the on-mass-shell, free-nucleon dispersion relation

$$p^0 = \sqrt{M^2 + \vec{p}^2} \quad , \tag{34}$$

i.e., an infinite Fermi gas of non-interacting nucleons, Eq. 31 correctly Fermi-averages the momentum distribution $n(\vec{p})d\vec{p}$. But of course the (trivial) Fermi gas model makes no reference to binding corrections.

To discuss bound nucleons in a realistic finite nucleus, we must abandon, first of all, the presumption that p^0 and $\vec{p}$ can be simultaneously specified. Elementary quantum mechanics tells us that, as they do not commute[11], they cannot both be specified. Since we cannot specify the 4-vector p of a bound nucleon, the Lorentz invariant inner product $p \cdot q$ no longer has any meaning. In other words, even though Eq. 30 is pleasing in form, we must abandon it for the more fundamental Eq. 28. The latter does not require anything so unphysical as the 4-momentum of a bound particle; all we need to know is a velocity pertaining to a (moving) bound nucleon, and the distribution of such velocities for the ensemble representing a bound nucleus.

The distinction between Eq. 28, which I believe to be correct in an appropriate semiclassical sense (to be defined), and Eq. 30 (which I am sure is wrong) has practical consequences. The reason is this: if we use the simple Doppler formula Eq. 28 we find

$$<F_2(x)> = \int d\vec{p}\, n(\vec{p}) F_2(x') \approx \int d\vec{p}\, n(\vec{p}) F_2\left(\frac{x}{\gamma(1 - \vec{v}\cdot\hat{q})}\right) \quad ; \tag{35}$$

whereas with Eq. 30,

$$<F_2(x)> = \int d\vec{p}n(\vec{p})F_2(x') \approx \int d\vec{p}n(\vec{p})F_2\left[\frac{x}{(p^0/M)(1 - \hat{q}\cdot\vec{v})}\right] . \quad (36)$$

Either Eq. 35 or 36 can be put in the convolution form of light-front dynamics[12]:

$$<F_2(x)> = \int dy f_N(y)F_2(x/y) \quad (37)$$

where Eq. 35 leads to

$$f_N(y) = \int d\vec{p}n(\vec{p})\delta\left[y - \gamma(1 - \vec{v}\cdot\hat{q})\right] . \quad (38)$$

A (nuclear) structure function $f_N(y)$ corresponding to Eq. 36 can be found with as little effort. This makes clear that light-front dynamics, *per se*, has little to say about nuclear structure, since as we shall now see, the consequences of Eq. 35 and 36 for bound nucleons are strikingly different.

In order to give Eq. 36 a meaning, we must assume something for p^0. Various prescriptions have been applied. For example, we can evaluate Eq. 38 state by state, for all the bound nucleons, and take

$$p^0 = M - B_k \quad (39)$$

where B_k is the binding energy of the k'th state. Or, we can take some form like

$$p^0 \approx M + \frac{\vec{p}^2}{2M} - \bar{V} \quad (40)$$

where $\bar{V}$ is some average potential[13]. But as long as we assume p^0 represents the single-particle energy of a nucleon, it is necessarily true that p^0/M is *smaller* than 1 (otherwise the nucleon would not be bound). Contrast this with the corresponding formula based on the simple Doppler effect, Eq. 35: here, instead of p^0/M we have

$$\gamma = \left[1 - \vec{v}^2 \right]^{-1/2}$$

which is *larger* than 1 as long as the velocity is real. All the binding effects are subsumed into the momentum distribution and the (semiclassical) relation between (group) velocity $\vec{v}$ and 3-momentum $\vec{p}$, that follows from regarding nucleons as wave packets within the nucleus.

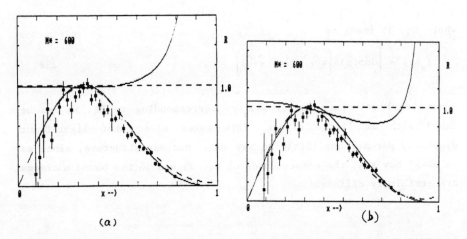

(a)   (b)

Fig. 5   The ratio $F_2^A(x)/F_2^d(x)$ using: a) the Doppler prescription
          b) the "standard" prescription.

Figure 5a,b above shows the widely accepted conclusion, that binding effects account for the EMC effect, arises from (erroneously) assuming naive Lorentz invariance for off-shell particles.

4. <u>What</u> <u>is</u> <u>the</u> <u>EMC</u> <u>Effect?</u>

If the EMC effect is not a binding effect, then what <u>is</u> it? Close, *et al.*[14] have attempted to exploit the (known) Q^2 dependence of structure functions by using a form of the Altarelli-Parisi rescaling equations. The problem with their theory is first, it is not a theory; and second, it should not apply equally to both the Stanford (low-Q^2) and CERN (high-Q^2) data, both of which exhibit the EMC effect. Finally, Thomas[15] has shown that — as I had initially suspected — the Q^2 dependence from gluonic bremstrahlung is too weak to explain the results.

Some years ago, in connection with quasielastic electron-nucleus scattering, I proposed that the elastic form factor of the proton softens in nuclear matter, compared with that of a free proton[16]. I did not at that time attempt to explain why this should happen, but noted that the softening was equivalent to an increase in the confinement radius of the quarks comprising the nucleon, by some 20-30%. Clearly, an increase of R will also soften the valence quark momentum distribution.

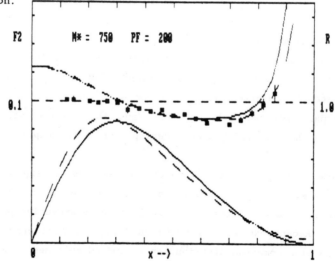

Fig. 6   The ratio of the non-singlet structure functions of Fe and ^2H, based on modifiying the confinement radius.

As Fig. 6 above shows, this is just what is needed to account for the EMC data at larger x. From the above discussion on the "singlet" part of $F_2(x)$, we may conclude that probably the glue momentum distribution softens also, giving results that seem qualitatively like those reported at low x by the NM Collaboration. I plan to check this in more detail.

5. Relation to the origin of mass

We presently believe quark confinement in QCD results from the structure of the vacuum, perhaps along the lines suggested by Parisi[17]. The details are unknown. Lattice gauge theory suggests that a deconfining phase transition occurs at a temperature ≈ 200 MeV. So far such temperatures have not been reached in, e.g., ultrarelativistic heavy-ion collisions. The nature of the condensate must be understood before we can understand the origin of the hadron masses. These are determined by a string tension parameter, which is related to the binding-energy density of the condensate.

The idea that the vacuum is a color-magnetic superconductor makes it possible to understand — with some modest asumptions, of course! — why nucleons expand in nuclei[18]. Flux tubes of finite volume and zero net color charge form because the total energy of color-electric field and broken vacuum must be a minimum. But once vacuum has been penetrated by color-electric flux, no more energy is required to let additional flux pass through it. Thus the average string tension decreases when several hadrons overlap. Hence they expand.

Such ideas suggest we can investigate confinement experimentally by using nuclei of varying density to "tune" the confinement radius. The experiments one would like to do would involve exclusive processes, of course. Recent exclusive results on J/ψ production provide evidence that the glue momentum distribution is indeed softened in a nucleus[19]. One may hope that by such means we can penetrate the mystery of mass.

6. Acknowledgements

I am grateful to Queen's University for their hospitality, to the EMC for sharing their data with me, and to C.H. Llewellyn Smith for several fruitful discussions.

REFERENCES

[1] J.J. Aubert, et al. (European Muon Collaboration), Phys. Lett. 123B (1983) 275.

[2] A. Bodek, et al., Phys. Rev. Lett. 50 (1983) 1431.
R.G. Arnold, et al., Phys. Rev. Lett. 52 (1983) 727.

[3] G. Bari, et al. (Bologna-CERN-Dubna-Munich-Saclay Collaboration), Phys. Lett. 163B (1985) 282.

[4] C.G. Callan and D.J. Gross, Phys. Rev. Lett. 22 (1969) 156.

[5] R.P. Feynman, Phys. Rev. Lett. 23 (1969) 1415;
J.D. Bjorken and E.A. Paschos, Phys. Rev. 185 (1969) 1975.

[6] In fact, the interactions must be ignored, since the energy and 3-momentum of a bound constituent cannot be specified simultaneously (they do not commute).

[7] R.L. Jaffe, Phys. Rev. D11 (1975) 1953.

[8] Called so because it can be considered photon absorption on virtual $q\bar{q}$ pairs in a hadron.

[9] J.V. Noble, Phys. Lett.

[10] S.K. Akulinichev, S. Shlomo, S.A. Kulagin and G.M. Vagradov, Phys. Rev. Lett. 55 (1985) 2239.

[11] Consider, e.g., a particle in a potential well: its energy depends on position and so cannot commute with its 3-momentum.

[12] C.H. Llewellyn-Smith, Phys. Lett. 128B (1983) 107.

[13]The best job is done by G.L. Li, K.F. Liu and G.E. Brown, "Role of Nuclear Binding in EMC Effect", U. of Kentucky preprint UK/88-01.

[14]F.E. Close, *et al.*, Phys. Lett. **129B** (1983); Phys. Rev. **D31** (1985) 1004.

[15]G. Dunne and A.W. Thomas, Nucl. Phys. **A446** (1985) 437c.

[16]J.V. Noble, Phys. Rev. Lett. **46** (1981) 412.

[17]G. Parisi, Phys. Rev. **D11** (1975) 970.

[18]J.V. Noble, Phys. Lett. **178B** (1986) 285.

[19]S. Katsanevas, *et al.*, Phys. Rev. Lett. **60** (1988) 2121.

GLUON INTERACTIONS AND PROTON SCATTERING

B. Margolis, P. Valin
Dept. of Physics, McGill University
3600 University St., Montreal H3A 2T8, Canada

and

F. Halzen
Dept. of Physics, University of Wisconsin
Madison, Wisconsin, 53706, USA

ABSTRACT

For values of $\sqrt{s} \approx$ few hundred GeV and greater, typical hadron collisions are semi-hard and calculable with QCD. Cross sections increase rapidly due to the strong increase of the gluon content of the proton. This results in an increasing pp and $\bar{p}p$ total cross section and a relatively large ratio of the real to imaginary forward scattering amplitude.

INTRODUCTION

There is renewed interest in the problem of hadron-hadron scattering because of recent higher energy experiments at the CERN $Sp\bar{p}S$ Collider and the Tevatron at Fermilab. We have been involved for some years in the description of high energy scattering and related phenomena using parton considerations[1]. The new data has given us clues which have made for a more precise description of these phenomena.

The new experiments include the measurement of $\sigma_{tot}(\bar{p}p)$, minijet production, the differential cross section for elastic scattering, $d\sigma_{el}/dt$ and hence $B(0)$ the forward elastic scattering slope and $\rho(0)$, the ratio of the real to imaginary forward scattering amplitude. The CERN UA4 Group[2] gives a large value $\rho(0) = 0.24 \pm .04$ at $\sqrt{s} = 546$ GeV.

The observation of a large value of $\rho(0)$ is a challenge to theory and model building. It can be explained using QCD as follows. For $\sqrt{s} \approx 1$ TeV typical hadron interactions become semi-hard and hence can be calculated in perturbative QCD. This is due to a rapid increase of the gluon content of the proton with increasing energy. This essentially threshold type behaviour results in a rapidly increasing total cross section and a relatively large value of $\rho(0)$. The proton in some sense becomes a "gluon bomb". This threshold onset has been seen by UA1 through the discovery of the minijet phenomena[3] (Figure 1). Certain cosmic ray measurements also indicate jet-like structures produced in Multi-TeV interactions[4].

GLUONS AND CENTRAL PRODUCTION OF HADRONS

Before the discovery of copious production of minijets at the $Sp\bar{p}S$ the dominant physical role of gluons in central hadron-hadron production was already evidenced by a characteristic scaling property[5]. Hadrons having quark structure similar to the proton are produced readily by fragmentation in proton-proton interactions whereas those of different quark composition like mesons and antibaryons are made centrally in the main. Figure 2 shows a plot of $M^3\sigma(M,s)/\Gamma$ plotted versus s/M^2, where $\sigma(M,s)$ is the inclusive cross section for making a resonance of mass M and width Γ in p-p interactions. This quantity has the behaviour

$$\sigma(M,s) = f(M) \, F_{gg}(M^2/s) \qquad (1)$$

where F_{gg} is the gluon-gluon structure function. We show in figure 2 two parallel curves proportional to the naive gluon-gluon structure function with n=5

$$F_{gg}(\tau) = 9 \int_\tau^1 \frac{dx}{x}(1-x)^5(1-\tau/x)^5 \qquad (2)$$

This scaling property is not dependent on the quark content of the produced resonance. Particles produced by fragmentation such as Δ^{++}, have cross sections that behave differently (see figure 2).

We show in figure 3 calculations of a simple model[6] of $\sigma(m,s)$ versus m at fixed $\sqrt{s}$ for several s for particles produced centrally. The model has as initiating mechanism gluon-gluon scattering followed by hadronization of the gluons. Also shown is experimental data (divided by $2J+1$). The cross section is determined mainly by the mass of the state, independent of quark content. For high enough mass at a given s the cross section becomes flatter as a function of mass. This is the region where low order perturbation contributions should be dominant.

It has been shown recently[7] that B meson production (i.e. b quark production) at $\sqrt{s} = 630$ GeV has as large an α_s^3 contribution as the lowest order α_s^2 gluon-gluon fusion result. The α_s^3 contribution is due to gluon-gluon scattering followed by gluon fragmentation into a $b\bar{b}$ pair. The measured B meson production cross section as well as charm production are better fit when the α_s^3 contribution is included. The centrally produced particles discussed above are in any case controlled by the gluon-gluon structure function. We demonstrate in the following that the dominant mechanism for particle production at high enough energy is gluon-gluon scattering followed by fragmentation of the gluons.

GLUONS, GLUON JETS AND DIFFRACTION

At high enough energies (e.g. $Sp\bar{p}S$) and moderate p_t the produced jets are gluon jets. The perturbative result for the jet cross section with transverse momentum p_t is valid not only in the hard scattering regime with $x_t = 2p_t/\sqrt{s} \approx 1$ but is in fact reliable for $p_t << \sqrt{s}$ when $p_t > \Lambda_{QCD}$. The cross section for jet production with $p_t > 1$ GeV through the dominant process $gg \to gg$ is

$$< n_g > \sigma_{tot} = \int_{1GeV} dp_t^2 \int dx_1 dx_2 g(x_1) g(x_2) \frac{d\hat{\sigma}}{dp_t^2} \qquad (3)$$

with

$$\frac{d\hat{\sigma}}{dp_t^2}(gg \to gg) \approx \frac{9\pi}{2} \frac{\alpha_s^2}{p_t^4} \qquad (4)$$

Using the naive structure function $g(x) = 3(1-x)^5/x$ and $\alpha_s \approx 0.2$ we find at $\sqrt{s} \approx 630$ GeV

$$< n_g > \sigma_{tot} \approx 80 \; mb \qquad (5)$$

The jet cross section is of the order of the total cross section and therefore at high energy where the gluon structure is fully developed hadron collisions become perturbative or semi-hard. The fact that a 1 GeV gluon jet cannot be resolved by experiment is irrelevant. The situation is analogous to that where the ratio of hadron production to muon yield in e^+e^- collisions

$$R = 3 \; \Sigma e_q^2 \qquad (6)$$

is valid already at $\sqrt{s} = 2$ GeV whereas jet structure of events does not become apparent until $\sqrt{s} \geq 7$ GeV.

The jet-like behaviour of hadron-hadron interactions appeared first in the laboratory in the minijet data of the UA1 experiment[3]. It becomes a feature of average interactions above 100 TeV apparent to Cosmic Ray detection[4]. This qualitative change in the event structure is responsible for violations of Feynman and KNO scaling in the energy range $\sqrt{s} = 0.1$ to 1 TeV. Why does this phenomenon appear to look like a threshold effect? This behaviour is associated with the rapid rise with energy of the number of relatively soft gluons once they take over the role of the valence quarks which control the behaviour of cross sections at lower energy ($\sqrt{s} \approx 100$ GeV). In the semi-hard regime the fractional gluon momenta x range from 10^{-2} to 10^{-3} typically and the gluon structure function increases faster than any power of x. A solution of the Altarelli-Parisi equations yields

$$xg(x, Q^2) \approx \exp\left(2\sqrt{\frac{3}{\pi b} \ln\left(\frac{\ln Q^2/\Lambda^2}{\ln Q_0^2/\Lambda^2}\right) \ln\frac{1}{x}}\right) \qquad (7)$$

with $b = (33 - 2n_f)/12\pi$ where n_f is the number of quark flavors. The threshold behaviour is enhanced further by the rapid evolution of g(x) with Q^2 in the small x region. This explosive small x behaviour is a feature of explicit structure functions such as EHLQ[8] and DO[9]. Figure 4 shows several forms of momentum distributions of gluons, $xg(x, Q^2)$, versus x for fixed Q^2 and several Q^2-independent structure functions.

We understand how perturbation theory breaks down when p_t approaches Λ_{QCD}. Screening of the large number of soft partons packed inside a high energy hadron modifies eq. (3) when x or p_t at a fixed energy becomes too small. This problem can be treated by introducing an eikonal formalism[1].

GLUON EIKONALIZATION

We calculate the total hadron-hadron cross section by constructing an elastic scattering amplitude and applying the optical theorem. The imaginary part of the scattering amplitude is

$$Im f(s, t) = \int_0^\infty (1 - e^{i\chi(s,b)}) J_0(b\sqrt{-t}) b \, db \qquad (8)$$

where the eikonal $\chi(s,b)$ is given by

$$\chi(s,b) = \frac{i}{2}W(b)\int_{m_0/\sqrt{s}}^{1}dx_1\int_{x_1}^{1}dx_2\int_{\delta^2}^{\hat{s}-\delta^2}d\hat{t}\,\frac{d\hat{\sigma}}{d\hat{t}}\,g(x_1,\hat{t})\,g(x_2,\hat{t}) \quad (9)$$

provided the glue-glue interaction is the only parton contribution. The parameters m_0 and δ remove the divergences of $d\hat{\sigma}/d\hat{t}$. We have here simply the QCD 2-jet cross section multiplied by a form factor $W(b)$ representing the gluon distribution in impact parameter space. We use the Fourier transform of the convoluted dipole form factors of the pp (or $\bar{p}p$) system to represent W(b)

$$W(b) = \frac{1}{8}(\mu b)^3 K_3(\mu b) \quad (10)$$

To be more realistic in terms of fitting scattering data from $\sqrt{s} \approx 20$ GeV and up we have considered more detailed models.

The following provides good fits to elastic scattering and total cross sections including the forward real-to-imaginary amplitude ratio, $d\sigma_{el}/dt$ and the forward elastic slope as a function of energy. We consider now an eikonal

$$\chi = \chi_{soft} + \chi_{gg} \quad (11)$$

where χ_{gg} is as in eq. (9) above and χ_{soft} represents soft processes and is taken to be asymptotically energy independent[1]. At the low energy end we have a small contribution to χ_{soft} from Regge pole terms[1]. Both contributions to χ are dependent on b through a density function of the form (10), possibly with different μ values for each term. We take the gluon distribution function

$$g(x) \approx \frac{1}{x^{1+\epsilon}}(1-x)^5 \quad (12)$$

The fits to the pp and $\bar{p}p$ data are shown in figures 5 through 8. The only parameters are the μ values, the constant describing $\chi_{soft} \approx const.$ $W(b)$, the cutoffs m_0 and δ (taken equal here), Λ_{QCD} and ϵ. All of these can be deduced from other data except μ for the gluons and ϵ. We find $\epsilon \approx .1$, corresponding to the behaviour of evolved structure functions for $Q^2 \approx 10$ GeV^2 as shown in figure 4. The value of μ for the gluons comes out to be slightly smaller than μ for χ_{soft} which is taken from the electromagnetic form factor of the proton. The real part[1] obeys

$$Ref(s,t) \approx \frac{\pi}{2}\frac{d}{dlns}Imf(s,t) \quad (13)$$

The fits shown in figure 5 yield $\rho(0) \approx .2$ at $\sqrt{s} = 546$ GeV. There is no problem in accounting for larger ρ values than predicted by previous calculations. The total cross section for $\epsilon > 0$ goes as $\sigma_{tot} \approx \log^2 s$ as $s \to \infty$ whereas for $\epsilon = 0$, $\sigma_{tot} \approx \log^2(\log s)$. The detailed form of $\hat{\sigma}_{gg}$ being slowly varying is of secondary importance. Here we have taken $\hat{\sigma}_{gg}(\hat{s}) = const. \; \theta(\hat{s} - \delta^2)$.

To calculate with confidence at very high energies, and for all values of t, one probably has to know more than we do. Certainly as we go to higher and higher energies the structure functions at smaller and smaller x come into play and these are not quantitatively established.

REFERENCES

1) Afek, Y., Leroy, C., Margolis, B. and Valin, P., Phys. Rev. Lett. 45, 85 (1980); L'Heureux, P., Margolis, B. and Valin, P., Phys. Rev. D32, 1681 (1985); Margolis, B., Valin, P. and Block, M.M., "High Energy Scenarios from Constituent Scattering", Proceedings of the IInd International Conference on Elastic and Diffractive Scattering (Rockefeller U., Oct 15-18 1987), R.L. Cool, K. Goulianos and N.N. Khuri eds.; Margolis, B., Valin, P., Block, M.M., Halzen, F. and Fletcher, R.S., "Forward Scattering Amplitudes in Semi-hard QCD", Madison Preprint MAD/PH/425 (May 1988), to appear in Physics Letters B.

2) UA4 Collaboration, Phys. Lett. 198B, 583 (1987).

3) UA1 Collaboration CERN preprint EP 88-29 (1988); Jacob, M. and Landshoff, P.V., Mod. Phys. Lett. A1, 657 (1986).

4) Yamdagni, N., in "Multiparticle Dynamics 1984", G. Gustafson and C. Peterson eds. (World Scientific, Singapore, 1984); Pancheri, G. and Srivastava, Y.N., and Kim, C.S. et al. in "Physics Simulations at High Energy", V. Barger et al., eds. (World Scientific, Singapore, 1986).

5) L'Heureux, P. and Margolis, B.,"Parton Picture of Particle Production", Proceedings of the 4th Annual MRST Meeting (McGill U., May 6-7 1982), B. Margolis and T.F. Morris eds.; Gaisser, T.K., Halzen, F. and Paschos, E., Phys. Rev. D15 2572 (1977); Halzen, F. and Matsuda, S., Phys. Rev. D17, 1344 (1978).

6) L'Heureux, P. and Margolis, B., Phys. Rev. D28, 242 (1983).

7) Nason, P., Dawson, S. and Ellis, R.K., Fermilab preprint Pub-87/222-T (1987); Altarelli, G., Diemoz, M., Martinelli, G. and Nason, P., CERN Preprint TH 4978/88.

8) Eichten, E.J., Hinchliffe, I., Lane, K. and Quigg, C., Rev. Mod. Phys. 56, 579 (1984).

9) Duke, D.W. and Owens, J.F., Phys. Rev. D30, 49 (1984).

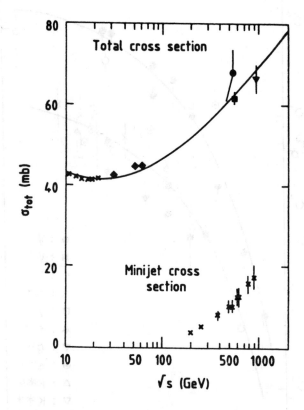

Figure 1: Measurements by the UA1 Collaboration[3)] of the cross section for events containing minijets with $p_t > 5$ GeV and $|\eta| < 1.5$, together with data for the total cross section.

186

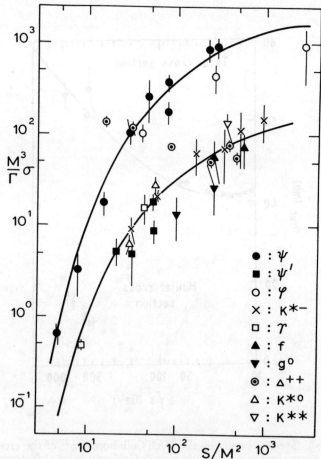

Figure 2: The quantity $M^3\sigma(M,s)/\Gamma$ vs s/M^2, where M is the resonance mass, $\sigma(M,s)$ the production cross section of M in pp interaction at center-of-mass energy $\sqrt{s}$ and Γ the total width of M. For references to data, see reference 1.

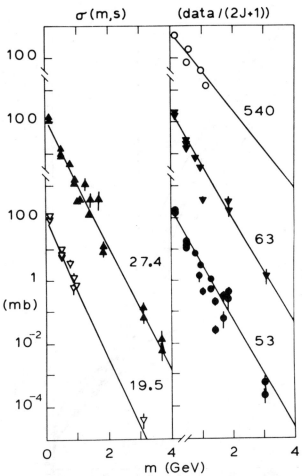

Figure 3: Inclusive single-particle cross section as a function of the mass m of the produced particle at five values of $\sqrt{s}$. For references to data, see reference 6.

188

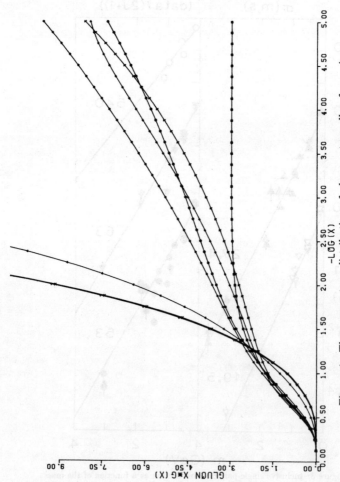

Figure 4: The momentum distribution of gluons at small x for various evolved and non-evolved parametrizations. Labelling the curves by their values at $x = 10^{-5}$ we have in counterclockwise ordering: the naive parton model with n=5, eqn. (12) with $\epsilon = .1$, EHLQ[8] Set 2, DO[9] "soft", EHLQ Set 1, the latter three being evaluated at $Q^2 = 10$ GeV^2, eqn. (12) with $\epsilon = .5$ and both EHLQs at $Q^2 = 546^2$ GeV^2, which almost coincide.

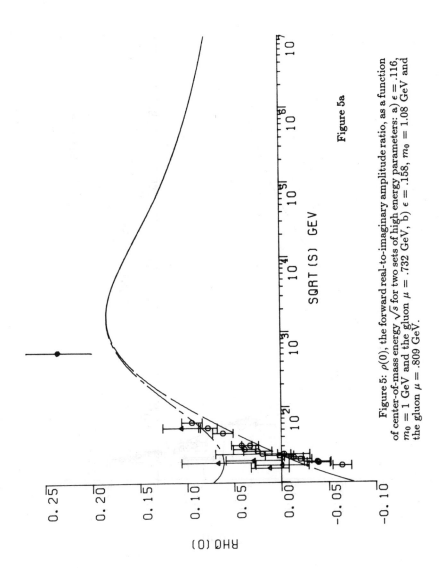

Figure 5: $\rho(0)$, the forward real-to-imaginary amplitude ratio, as a function of center-of-mass energy $\sqrt{s}$ for two sets of high energy parameters: a) $\epsilon = .116$, $m_0 = 1$ GeV and the gluon $\mu = .732$ GeV, b) $\epsilon = .158$, $m_0 = 1.08$ GeV and the gluon $\mu = .809$ GeV.

Figure 5a

190

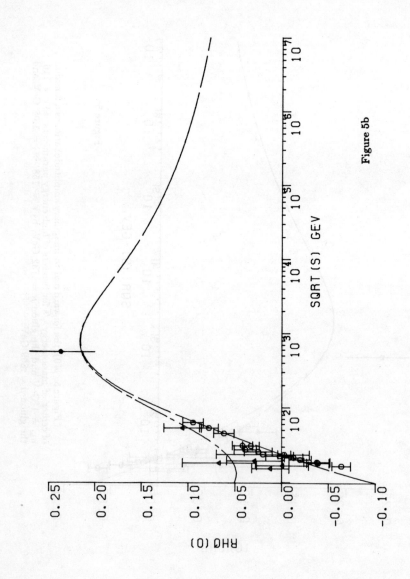

Figure 5b

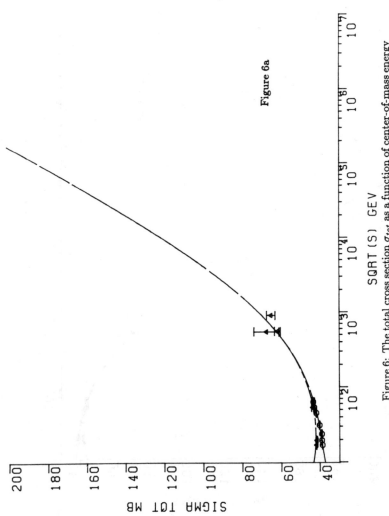

Figure 6a

SQRT (S) GEV

SIGMA TOT MB

Figure 6: The total cross section σ_{tot} as a function of center-of-mass energy $\sqrt{s}$ for the same sets of parameters a) and b) as in Figure 5.

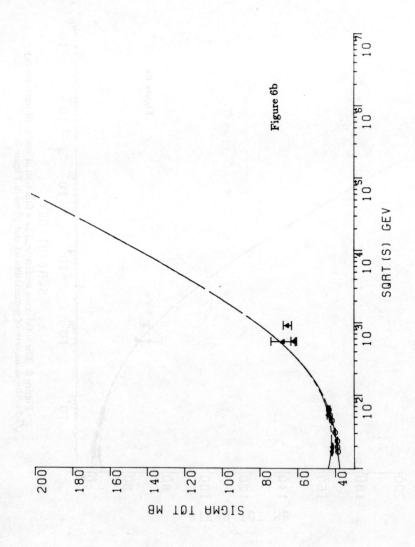

Figure 6b

193

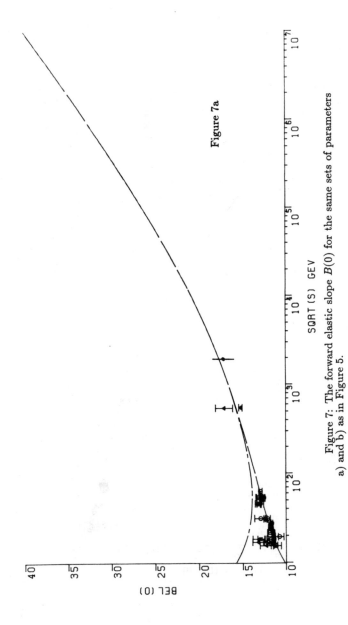

Figure 7a

SQRT(S) GEV

BEL(0)

Figure 7: The forward elastic slope $B(0)$ for the same sets of parameters
a) and b) as in Figure 5.

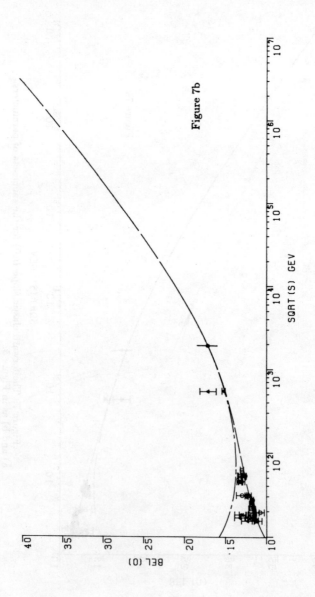

Figure 7b

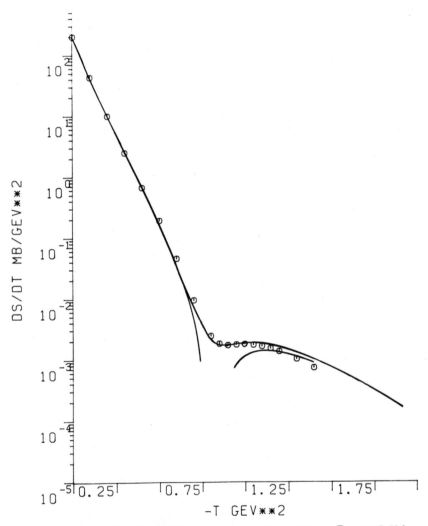

Figure 8: The elastic differential cross section $d\sigma/dt$ at $\sqrt{s} = 546$ GeV for parameter set a) of Figure 5.

Electroweak Interactions

"STRONG CORRECTIONS TO EFFECTIVE

OPERATORS OF b TO s PROCESSES"

ROBERT GRIGJANIS[†]

Department of Physics

University of Toronto, Toronto, Ontario, Canada M5S 1A7

ABSTRACT

It appears that rare decays of the B meson may soon be detectable. At lowest order, decays such as the inclusive $B \to X_s, \gamma$ are suppressed by the GIM mechanism[1]. However, disparate mass scales involved in the process raise the possibility of significant logarithmic QCD corrections. The results obtained here exhibit such an enhancement, but disagree with previous authors.

† Research supported by NSERC and NATO (grant 248/88)

INTRODUCTION

The importance of strange decays of the B meson from a theoretical standpoint is their sensitivity to the mass of the top quark, which is an intermediate particle in these processes. For a sufficiently light top quark , the decays are strongly suppressed by the GIM mechanism. However, it appears that the top mass may be greater than 50 Gev[2], and thus comparable to the mass of another intermediate particle, the W vector boson. QCD corrections arising from this difference in mass scales (intermediate versus external) are expected to be on the order of $(\alpha_s(m_b)/\pi)\ell n(m_t/m_b))$ which is the same order of magnitude as the lowest-order amplitude for $B \to \chi_s \gamma$[3]. In order to evaluate these corrections, it is advantageous to make certain approximations. Firstly, the partner of the b quark in the B meson is treated as a spectator. This reduces the problem to the free decay of a b quark. Secondly, the differing mass scales allow the application of the large mass expansion[4], which eliminates (integrates out) heavy intermediate fields at the expense of introducing effective local operators. Physical amplitudes can be directly related to the co-efficients of these operators, whose renormalization group behaviour is obtained in a straightforward manner. A third approximation is to neglect contributions from the u and d quarks. This depends on an assumption about the elements of the CKM quark-mixing matrix V_{ij}[5], which occurs in the coupling of the quarks to the W boson;

$$V_{ub}V_{us}^* << V_{cb}V_{cs}^* \simeq -V_{tb}V_{ts}^* \tag{1}$$

Thus far, we have followed in the path of previous authors[6]. However, an important technical point that has been ignored is the occurrence of γ_5 matrices in Feynman diagrams; this renders the use of dimensional regularization invalid unless modifications are introduced. The method used here is dimensional reduction[7], which, simply put, has all space-time indices running from 0 to 3 except those of momenta (or spacetime coordinates), which run from 0 to 3-ϵ. This prescription appears to cure γ_5 problems as long as there are no closed fermion loops (e.g. - axial anomaly). Since no such traces occur in the work presented here, the applicability of dimensional reduction is at worst more plausible than that of naive dimensional regularization. It should be noted that, apart from the corrections arising from this technique, the results obtained here agree with those of Grinstein et al[6].

Section 1 discusses the lowest-order effective Lagrangian for b→s processes. Section 2 introduces QCD corrections, and the resulting anomalous dimension matrix for the operator co-efficients. Section 3 presents the solutions to the renormalization group equations, and thus the leading logarithm solution for the amplitude b→s. Finally, the branching ratio for B→ $X_{,s}\gamma$ is given.

The Effective Lagrangian At Lowest Order

In the standard model SU(3) colour × (SU(2) × U(1)) electroweak with a single complex Higgs doublet, the light particles are the s,c,b quarks, the photon, and the gluon. The heavy particles relevant to b→s processes are the W vector boson, the charged Higgs scalar $\phi^{\pm}$ ($M_{\phi} = M_w$ in the Feynman gauge with gauge parameter ξ=1), and the top quark. The first step in obtaining the effective Lagrangian is to compute one-light-particle-irreducible (ILPI) Feynman diagrams with only light external particles, and at least one internal heavy particle. ILPI means that the diagram becomes proper (IPI) if the heavy propagator (s) is (are) contracted to a point. Using the background-field gauge fixing terms for W[4], the resulting functions of external momenta can be viewed as components of gauge-invariant local operators. Only operators of dimension five and six are retained; those of lower dimension can be absorbed by redifinition of masses, couplings and fields, while operators of higher dimension are suppressed by powers of (external momentum/M_w). Because of gauge invariance, one need only consider diagrams with (a) external particles b,s, and one photon; (b) external particles b,s and one gluon; (c) the four-quark diagrams with external quarks b,s,c,c. The relevant diagrams are shown in Figure 1.

Using dimensional regularization with dimensional reduction, one encounters factors of $\ell n(M_w/\mu)$, where μ is the regularization mass scale. To ensure the validity of the perturbation expansion, the scale must be chosen such that these logarithms are small. The most obvious choice is $\mu = M_w$. Thus, all couplings, masses and operator co-efficients are defined at this scale. However, the physical scale is the mass of the b quark (or the B meson), so that renormalization group techniques must be used to obtain physical amplitudes, with the lowest-order parameters determining the inital conditions. Given the mass scales (M_w=82 GeV; m_b =5 GeV), the obvious candidate for the scaling mechanism is QCD corrections.

To continue at lowest order; from the momentum expansions of the diagrams in Fig. 1, nine gauge-invariant operators of dimension five or six can be identified. In the following, $D_\mu = \partial_\mu - ieQ_dA_\mu - ig_st_aG^a\mu$, where eQ_d is the charge of the down quark, $A_\mu(G^a_\mu)$ is the photon (gluon) field, g_s is the strong coupling constant, and $t_a(a = 1, .., 8)$ are the hermitian generators of SU(3) colour in the fundamental representation (normalized to $\text{TR}(t_at_b) = \frac{1}{2}$). S_L denotes the left-handed component of the S field; $S_L = P_LS$, where $P_L = \frac{1}{2}(1 - \gamma_5)$. $F_{\mu\nu}(G^a_{\mu\nu})$ denotes the electromagnetic (strong) field strength. In O_9, α and β are colour indices.

$$O_1 = \frac{1}{(4\pi)^2}i\bar{s}(\not{D})^3b_L \tag{2}$$

$$O_2 = \frac{1}{(4\pi)^2}ieQ_d\bar{s}_L\left\{\not{D}, F^{\mu\nu}\sigma_{\mu\nu}\right\}b_L \tag{3}$$

$$O_3 = \frac{1}{(4\pi)^2}eQ_d\bar{s}_L\gamma_\nu b_L\partial_\mu F^{\mu\nu} \tag{4}$$

$$O_4 = \frac{1}{(4\pi)^2}\bar{s}(m_bP_R + m_sP_L)(\not{D})^2b \tag{5}$$

$$O_5 = \frac{1}{(4\pi)^2}eQ_d\bar{s}(m_bP_R + m_sP_L)F^{\mu\nu}\sigma_{\mu\nu}b \tag{6}$$

$$O_6 = \frac{1}{(4\pi)^2}ig_s\bar{s}_L\left\{\not{D}, t_aG^a_{\mu\nu}\sigma^{\mu\nu}\right\}b_L \tag{7}$$

$$O_7 = \frac{1}{(4\pi)^2}g_s\bar{s}_L\gamma_\nu t_ab_L(D_\mu G^{\mu\nu})_a \tag{8}$$

$$O_8 = \frac{1}{(4\pi)^2}g_s(m_bP_R + m_sP_L)t_aG^{\mu\nu}_a\sigma_{\mu\nu}b \tag{9}$$

$$O_9 = (\bar{s}_{L\alpha}\gamma^\mu c_\alpha)(\bar{c}_\beta\gamma_\mu b_{L\beta}) \tag{10}$$

The momentum dependence of these operators for b(p)→s(p')+ gauge boson (q) is given below, with ϵ^μ being the gauge boson polarization vector.

$$O_1 \qquad p^2\slashed{\epsilon} + p'2\slashed{\epsilon} + \slashed{p}'\slashed{p}$$

$$O_2, O_6 \qquad \slashed{p}'[\slashed{q}, \slashed{\epsilon}] + [\slashed{q}, \slashed{\epsilon}]D\slashed{p}$$

$$O_3, O_7 \qquad q^2\slashed{\epsilon} - \slashed{q}(q \cdot \epsilon)$$

$$O_4 \qquad (m_b P_R + m_s P_L)(\slashed{p}'\slashed{\epsilon} + \slashed{\epsilon}\slashed{p})$$

$$O_5, O_8 \qquad (m_b P_R + m_s P_L)[\slashed{q}, \slashed{\epsilon}]$$

The lowest-order effective Lagrangian can now be written;

$$\mathcal{L}_{eff} = \mathcal{L}_{SU(3) \times U(1)} + \frac{4G_f}{\sqrt{2}} V_{ts}^* V_{tb} \left(\sum_{i=1}^{7} C_i O_i + \mu^\epsilon C_9 O_9 \right) \qquad (11)$$

Here, G_f is the Fermi constant. The factor of μ^ϵ is necessary for dimensional reasons; $\dim \mathcal{L} = 4 - \epsilon$, $\dim G_F = -2$, $\dim C_i = 0$, $\dim O_9 = 6 - 2\epsilon$. The coefficients C_i are functions of the scale μ, and the ratio $\delta = m_t^2/M_w^2$. Ratios involving the light quarks instead of the t are neglected. The lowest-order calculations give initial conditions for the coefficients $C_i(\mu, \delta)$;[8]

$$C_i(M_w, \delta) = c_i(\delta) \qquad i = 1, ..., 8 \qquad c_9(M_w) = 1 \qquad (12)$$

The amplitude for the decay $b \to s\gamma$ is proportional to $C_2(m_b, \delta) + C_5(m_b, \delta)$, and it is to the computation of this that we now address ourselves.

2. QCD Corrections

The next step is to evaluate all proper $b \to s\gamma$, $b \to s+$ gluon and four-quark diagrams, involving one of the operators O_i, and an internal gluon. The momentum expansions of these diagrams determine QCD counterterms for the operators O_i. Firstly, one finds that these corrections induce operators not found at lowest order. Adding an internal gluon to the operator O_9 induces the new four-quark operator

$$O_{10} \equiv (\bar{s}_{L\alpha} \gamma^\mu c_\beta)(\bar{c}_\beta \gamma_\mu b_{L\alpha}) \qquad (13)$$

Its coefficient $C_{10}(\mu)$ is given the intial condition

$$C_{10}(M_w) = 0 \qquad (14)$$

In addition, certain non-gauge-invariant operators arise. Their presence is related to the non-Abelian nature of the SU(3) colour group, and the resulting non-linearity of the Ward identities. Nevertheless, they are easily isolated, and do not contribute to amplitudes when the b and s quarks are on shell. As an example of operator mixing, the QCD counterterms generated by the operator O_8 are given. The relevant diagrams are shown in Figure 2.

$$
C_8 O_8 \rightarrow C_8 \frac{\alpha_s}{2\pi} \epsilon \left\{ 8 O_4 + \frac{16}{3} O_5 - \frac{11}{3} O_8 \right.
$$

$$
+ \frac{9}{4(4\pi)^2} g_s \bar{s} t_a (m_b P_R + m_s P_L) \slashed{G}_a (i\slashed{\partial} - m_b) b
$$

$$
\left. + \frac{9}{4(4\pi)^2} g_s \bar{s} (i\slashed{\partial} - m_s) t_a \slashed{G}_0 (m_b P_R + m_s P_L) b \right\}
$$

Here, $\alpha_s = g_s^2 / 4\pi$. The last two terms are examples of the new operators mentioned above, and their irrelevance to on-shell amplitudes is apparent. They will not be referred to henceforth, through the importance of isolating them must be emphasized.

The b$\rightarrow$ s sector of the effective Lagrangian is now (at scale μ)

$$
\sum_{i=1}^{8} \sum_{j=1}^{10} O_i (C_i + \frac{\alpha_s}{2\pi\epsilon} \tilde{\gamma}_{ij} C_j)
$$

$$
+ \sum_{i,j=9}^{10} \mu^\epsilon O_i (C_i + \frac{\alpha_s}{2\pi\epsilon} \tilde{\gamma}_{ij} C_j)
$$

$$
= \sum_{i=1}^{10} O_{iB} C_{iB}
$$

$$
= \sum_{i=1}^{10} Z_i O_i C_{iB} \tag{15}
$$

Identification with the scale-independent bare Lagrangian has been made. The factors Z_i incorporate the mass, coupling and wavefunction renormalization constants relating bare to renormalized parameters in O_i. Scale-independence of C_{iB} gives

$$
\mu \frac{\partial}{\partial \mu} C_{iB} = O = \begin{cases} \mu \frac{\partial}{\partial \mu} \sum_{j=1}^{10} Z_i^{-1}(C_i + \frac{\alpha_s}{2\pi\epsilon} \tilde{\gamma}_{ij} C_j), i = 1,..,8 \\ \mu \frac{\partial}{\partial \mu} \sum_{j=9}^{10} Z_i^{-1} \mu^\epsilon (C_i + \frac{\alpha_s}{2\pi\epsilon} \tilde{\gamma}_{ij} C_j), i = 9, 10 \end{cases} \tag{16}
$$

Picking out terms of $O(\epsilon)$ and $O(1)$ gives the renormalization group equations;

$$\mu\frac{\partial C_i}{\partial \mu} = \frac{\alpha_s}{2\pi}\gamma_{ij}C_j \tag{17}$$

where γ_{ij} is the 10×10 anomalous dimension matrix. Fortunately, this matrix is not as formidable as may be expected. The operators corresponding to emission of real photons (O_2, O_5) mix only with each other, with the analogous gluon operators (O_6, O_8), and with the four-quark operators $(O_{9,10})$. That is to say, it is only these operators that generate counterterms for O_2 and O_5. So, to analyze b$\to$ sγ we need only the 6×6 sub-matrix $\hat{\gamma}_{ij}$ (i,j = 2, 5,6,8,9,10). If the counterterm for O_i (i=2,5,6,8) induced by the four-quark operator $O_9(O_{10})$ is

$$\frac{\alpha_s}{2\pi\epsilon}C_9 x_i \left(\frac{\alpha_s}{2\pi\epsilon}C_{10}y_i\right)$$

the the sub-matrix $\hat{\gamma}_{ij}$ is

$$\hat{\gamma}_{ij} = \begin{vmatrix} 0 & 0 & \frac{8}{3} & 0 & 2x_2 & 0 \\ \frac{16}{3} & \frac{16}{3} & \frac{8}{3} & \frac{16}{3} & 2x_5 & 0 \\ 0 & 0 & \frac{41}{12} & 0 & 2x_6 & 2y_6 \\ 0 & 0 & \frac{5}{4} & \frac{14}{3} & 2x_8 & 2y_8 \\ 0 & 0 & 0 & 0 & -1 & 3 \\ 0 & 0 & 0 & 0 & 3 & -1 \end{vmatrix}$$

The factors of 2 multiplying x_i, y_i arise essentially from the extra μ^ϵ multiplying the four-quark operators in $\mathcal{L}_{eff}$. These numbers come from the single pole $(1/\epsilon)$ parts of diagrams like the one shown in Fig. 3. Because of the two-loop structure of such diagrams, double poles occur $(1/\epsilon^2)$, and improper treatment of the γ_5 matrices (i.e. dimensional regularization with $\{\gamma_\mu, \gamma_5 = 0\}$) will lead to errors in x_i y_i.

Notice that by adding rows together, one can obtain equations in terms of the combinations;

$$\tilde{C}_5(\mu) = C_2(\mu) + C_5(\mu) \tag{21}$$

$$\tilde{C}_8(\mu) = C_6(\mu) + C_8(\mu) \tag{22}$$

And, with $X \equiv 2(x_6 + x_5), Y \equiv 2(x_6 + x_8), Z = 2(y_6 + y_8)$, these equations are

$$\mu\frac{\partial \tilde{C}_5}{\partial \mu} = \frac{\alpha_s}{2\pi}\left\{\frac{16}{3}\tilde{C}_5 + \frac{16}{3}\tilde{C}_8 + XC_9\right\}$$

$$\mu\frac{\partial \tilde{C}_8}{\partial \mu} = \frac{\alpha_s}{2\pi}\left\{\frac{14}{3}\tilde{C}_8 + YC_9 + ZC_{10}\right\}$$

Computation of the two-loop diagrams with four-quark operators in them yields

$$X = \frac{124}{27} \qquad Y = \frac{73}{108} \qquad Z = -1 \tag{23}$$

3. Solutions, and Branching Ratio

The only additional piece of information required to solve the renormalization group equations is that, in the region of integration ($m_b < \mu < M_w$), there are effectively five flavours of quark. If the top mass is less than that of the W, this will lead to an error (in the step-function approximation) proportional to $ln(m_t/M_w)$, which is presumed to be small.

With the notation $\eta \equiv \alpha_s(m_b)/\alpha_s(M_w)$ the solutions for $\tilde{C}_5, \tilde{C}_8$ are, in the leading logarithm approximation;

$$\tilde{C}_5(m_b) = \eta^{-\frac{16}{23}} \left\{ \tilde{C}_5(M_W) - \left[8\tilde{C}_8(M_W) + \frac{3}{26}(17Y + 9Z) \right] (\eta^{\frac{2}{23}} - 1) \tag{24} \right.$$

$$- \frac{3}{20}(X - 2Y - 2Z)(\eta^{\frac{10}{23}} - 1)$$

$$\left. - \frac{3}{56} \left[\frac{8}{13}(Z - Y) + X \right] (\eta^{\frac{28}{23}} - 1) \right\}$$

$$\tilde{C}_8(m_b) = \eta^{-\frac{14}{23}} \left\{ \tilde{C}_8(M_W) - \frac{3}{16}(Y + Z)(\eta^{\frac{8}{23}} - 1) \right.$$

$$\left. + \frac{3}{52}(Z - Y)(\eta^{\frac{26}{23}} - 1) \right\} \tag{25}$$

It now remains only to obtain a measurable number -i.e. a branching ratio. By far the dominant inclusive decay mode is B$\rightarrow X_c e\bar{\nu}_e$. Thus the branching ratio is

$$\frac{\Gamma(B \rightarrow \gamma X_s)}{\Gamma(B \rightarrow_c e\bar{\nu}_e)} = \frac{2\alpha|\tilde{C}_5(m_b)|^2}{3\pi\rho(m_c/m_b)} \frac{1}{(1 - \frac{2\alpha_s(m_b)}{3\pi f(m_c/m_b))})} \tag{26}$$

where α is the fine structure constant and the phase-space factor $\rho(m_c/m_b) \simeq 0.447$. Also shown in the denominator are the one-loop corrections[9] to the semi leptonic decay, with $f(m_c/m_b) \simeq 2.4$. The ratio is given in the table for a range of values for the top quark mass from 50 to 160 GeV and for Λ_{QCD} =100, 200 and 250 MeV.

From the table we can see the sensitivity of the ratio to variations in Λ_{QCD}. Since we have integrated out the top quark, taking the mass to be comparable to that of M_W, the error in the calculation increases as m_t decreases and is estimated to be of the order of 15% at $m_t \simeq 50$ GeV.

Conclusions

The main conclusions arising from the work presented here are: (a) Through the renormalization group equations, the b→s + gluon amplitude contributes to the b→ $s\gamma$ amplitude. (b) b→s + gluon is strongly suppressed by QCD corrections, and its contribution in (a) is negligible. (c) Thus, the correction to the branching ratio $\Gamma(B \to X_s\gamma)/\Gamma(B \to X_c e\bar{\nu}_e)$ depends mainly on the value of X, whose value here differs from that obtained by other authors, because of the use of dimensional reduction. (d) The resulting enhancement to the branching ratio ranges from about 7 at M=50 GeV, to about 2 at $m_t = 160$ GeV.

References

1. S.L. Glashow, J. Iliopoulos and L. Maiani, Phys. Rev. **D2** (1970) 1285.

2. H. Albrecht et al., Phys. Lett. B **192** (1987) 239; W. Schmidt-Parzefall, Nucl. Phys. B (Proc. Suppl.) 3 (1988) 253.

3. B.A. Campbell and P.J. O'Donnell, Phys. Rev. D25 (1982) 1989.

4. S. Weinberg, Phys. Lett. **B91** (1980) 51.

5. N. Cabibbo, Phys. Rev. Letters 10 (1963) 531; M. Kobayashi and T. Maskawa, Prog. Theor. Phys. 49 (1973).

6. B. Grinstein, R. Springer and M.B. Wise, Phys. Lett. **B202** (1988) 138.

7. G. Altarelli et. al., Nucl. Phys. **B187** (1981) 461.

8. R. Grigjanis et.al., University of Toronto preprint UTPT-88-11 (To appear in Phys. Lett. B).

9. N. Cabibbo and L. Maiani, Phys. Lett. **B79** (1978) 109.

Table Caption

For a number of values of the top quark mass, the ratio of the inclusive decay $b \to s\gamma$ relative to the semileptonic decay is shown, together with the dependence of the QCD corrections on Λ_{QCD}. The right hand columns show $|\text{amplitude}|^2$, both uncorrected and corrected by QCD, for the process $b \to sG$.

Mt (GeV)	$\frac{\Gamma(B \to X_s \gamma)}{\Gamma(B \to X_c e \bar{v}_e)} \times 10^4$				$\lvert \tilde{C}_8 \rvert^2 \times 10^4$			
	uncorrected	QCD corrected $\Lambda_{QCD}(MeV)$			uncorrected	QCD corrected $\Lambda_{QCD}(MeV)$		
		100	200	250		100	200	250
50	1.0	6.3	7.5	8.0	15	4.6	9.8	12.4
60	1.7	7.4	8.6	9.2	22	2.4	6.6	8.8
80	3.2	9.7	11.0	11.6	36	0.37	2.8	4.3
100	5.0	11.9	13.3	13.8	52	0.07	0.70	1.6
120	6.8	14.0	15.4	16.0	66	0.79	0.06	0.48
140	8.7	15.9	17.3	17.9	77	1.9	0.05	0.06
160	10.4	17.7	19.1	19.7	87	3.1	0.35	0.02

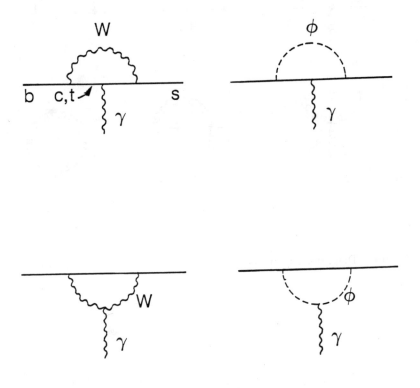

Figure 1: Diagrams yielding effective operators for b→ sγ at lowest order.

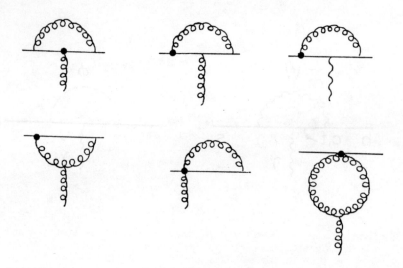

Figure 2: Diagrams leading to counterterms generated by the operator O_8

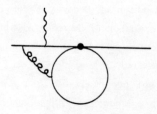

Figure 3: A diagram giving contributions from the four-quark operator O_9, to $b \rightarrow \gamma s$

CORRECTIONS TO SUPERALLOWED FERMI β-DECAY AND THE DETERMINATION OF V_{ud}

I.S. Towner

Chalk River Nuclear Laboratories, Atomic Energy of Canada Limited Research Company, Chalk River, Ontario, Canada K0J 1J0

Abstract

Data on superallowed Fermi β-decay in nuclei are surveyed and electromagnetic corrections discussed. Recommended values for the weak vector coupling constant and the leading element of the Kobayashi-Maskawa mixing matrix are given. Comments on pion β-decay and neutron decay are made.

1. INTRODUCTION

The intensity of superallowed Fermi β-transitions between 0^+ T=1 nuclear states as expressed through their ft values is expected to be the same for all nuclei. This statement follows from two tenets. First, in the allowed approximation of β-decay theory, the Fermi operator, τ_+, being just the ladder operator of isospin algebra, leads to identically-valued matrix elements for all Fermi decays so long as isospin is an exact symmetry. And second, the weak vector current, believed to be a conserved current, is not renormalised in a nuclear many-body medium by such processes as meson-exchange currents, and as a consequence the weak vector coupling constant is indeed a true constant. However, before these statements can be tested against experimental data, certain theoretical corrections have to be applied. For example, bremsstrahlung processes lead to radiative corrections. And the breakdown of analogue symmetry by the presence of charge-dependent forces between nucleons leads to Coulomb corrections in the nuclear matrix elements.

The last comprehensive survey[1] of the data was presented at the 1984 conference on atomic masses and fundamental constants. At that time there was a troublesome discrepancy between data on the low-Z nuclei (O, Aℓ, Cℓ, K) and the high-Z nuclei (Sc, V, Mn, Co). The corrected ft values in each group were internally consistent, but the average values for the two groups differed from each other by about five standard deviations. This unsatisfactory situation was resolved two years later through a re-evaluation[2] of the radiative correction of order $Z\alpha^2$. The discrepancy also prompted activity[3,4] on the other major theoretical correction associated with charge-dependent nuclear forces.

In this report we discuss these recent developments and update the 1984 survey.

2. RADIATIVE CORRECTIONS

The interaction of the charged particles, electrons or protons, with the external electromagnetic field leads to corrections of order $(Z\alpha)^m$, $m = 1, 2, \cdots$, where Z is the charge number for the daughter nucleus in the β-decay and α the fine-structure constant. For electrons, this correction is included in the evaluation of the phase space integral, f, by using exact solutions of the Dirac equation for electrons in the field of a modelled nuclear charge distribution rather than plane-wave electron wavefunctions. Similarly for protons, the correction is incorporated as a Coulomb correction to the nuclear matrix element to be discussed in the next section.

The remaining radiative corrections are of order $Z^m\alpha^n$ with $m < n$ and must be considered explicitly. The leading term is the order α correction due to the exchange of a photon (or a neutral Z-vector boson) between the electron and the proton. Sirlin[5] has evaluated this term and the result can be written as a sum of two pieces: an inner radiative correction, Δ_β, that does not depend on the electron energy (and hence not on the nucleus in question) but is sensitive to the modelling of the weak interaction process itself, and an outer radiative correction, δ_R. Since the test of constancy of ft values from one nucleus to another is not affected by the inner radiative correction, we put this aside for the moment and consider just δ_R. To order α, δ_R can be evaluated with standard quantum electrodynamics without ambiguity. The correction is typically of order 1%.

Higher order corrections of order $Z\alpha^2$ and $Z^2\alpha^3$ have, until recently, only been calculated by Jaus[6] in the early '70s and were estimated to grow rapidly with Z from 0.3% in ^{14}O to 1.2% in ^{54}Co. In this work, the order $Z\alpha^2$ correction has the structure

$$\delta_R(Z\alpha^2) \simeq Z\alpha^2\{\ln(M/m) + \Delta^\circ(E_o)\} \tag{1}$$

where the leading logarithm is obtained analytically and the second term evaluated numerically. Here M is the nucleon mass, m the electron mass and E_o the maximum electron energy in electron-mass units. Jaus found $\Delta^\circ(E_o)$ to be a positive and monotonically increasing function of E_o. Motivated by the experimental discrepancy noted in the introduction, Sirlin and Zucchini[2] sought an analytic way to estimate the second term and on treating the electron in an extreme relativistic approximation came to the form

$$\delta_R(Z\alpha^2) = Z\alpha^2\{\ln(M/m) - \ln(2E_o) + c + \cdots\} \tag{2}$$

showing that $\Delta^\circ(E_o)$ should be a monotonically decreasing function of E_o. Furthermore, unless c is a large positive constant, one expects $\Delta^\circ(E_o)$ to be negative especially for the high-Z cases. The determination of c requires a detailed calculation. This result, eq. (2), has been essentially confirmed in subsequent calculations by Jaus & Rasche[7] and Sirlin[8]. The higher order corrections increase from light to heavier nuclei, but slower than Z, and are computed to be 0.2% in ^{14}O and 0.6% in ^{54}Co. This re-evaluation removes the problematical discrepancy.

3. COULOMB CORRECTION

The other nuclear-structure dependent Coulomb correction is the most troublesome part of the analysis since it is heavily dependent on nuclear models. The presence of Coulomb and charge-dependent nuclear forces destroys the isospin symmetry, and the Fermi matrix elements between analogue T=1 states is modified: $\langle f|\tau_+|i\rangle^2 = 2(1-\delta_c)$. There are two physical phenomena that can contribute to δ_c. First, the degree of configuration mixing in the shell-model wavefunctions varies from member to member within an isospin multiplet, leading to a correction δ_{C1}. Second, protons are typically less bound than neutrons. Thus the tail of the radial wavefunction for protons extends further than that for neutrons, so the radial overlap of the parent and daughter nuclei is reduced from the normally assumed value of unity, a correction δ_{C2}. Generally $\delta_{C2} > \delta_{C1}$, so we will discuss this first.

3.1   The Correction, δ_{C2}

The correction, δ_{C2}, has been evaluated by Wilkinson[9] and Towner, Hardy and Harvey[10] and the two calculations are generally in agreement. The correction is of order 0.5%. The idea is that the proton in the parent nucleus undergoing β-decay and the neutron in the daughter nucleus can each be described by a one-body Schrodinger equation in which the nucleon is assumed to be moving in an average central potential due to all the other nucleons. The average central potential is modelled. It is usually assumed to be of Saxon-Woods form, with the depth of the potential adjusted so that the computed eigenvalue matches the experimentally known separation energy obtained from mass differences. The correction, δ_{C2}, is the departure from unity of the radial overlap integral of single-particle proton and neutron Saxon-Woods wavefunctions.

A refinement of the scheme acknowledges a many-body facet to this problem. The parent state for beta decay, say ^{14}O to take a specific example, is not just written as ^{13}N(g.s.) + p but is expanded in a complete set of ^{13}N states. The expansion coefficients can be estimated in a shell model calculation. This modification leads to a 30% enhancement in the calculated correction.

There still remains the practical detail of choosing the other parameters of the Saxon-Woods form, most notably the radius parameter. Towner, Hardy & Harvey[10] computed the charge radius of the nucleus from the proton eigenfunctions of the Saxon-Woods potential and adjusted the potential parameter so that the charge radius agreed with measurements from electron scattering. These authors also give an uncertainty in δ_{C2} arising from the uncertainty in the potential parameters coming from the experimental uncertainty in the charge radius. However, this error estimate can only be interpreted in the context of the Saxon-Woods model and makes no allowance for the correctness of the model *per se.*

A theoretically more satisfying approach is to obtain the average central potential from a self-consistent Hartree-Fock calculation. This has been attempted by Ormand & Brown[3,4]. A choice still has to be made for the bare nucleon-nucleon interaction. Ormand & Brown use five different variants of the Skyrme interaction known to give good results for binding energies and radii of closed shell nuclei in Hartree-Fock calculations. Following a self-consistent computation an average central potential is obtained whose overall strength is further adjusted so that the single-particle eigenvalues agree with experimental separation energies, as was done in the Saxon-Woods calculation. The departure from unity of the overlap of the single Slater determinant wavefunctions for the parent and daughter nucleus leads to a value of δ_{C2}. (It is important in this approach to overlap the entire determinant of A single-particle functions and not just the last nucleon functions as pointed out in ref. 4). The range of values obtained with the five different variants of the Skyrme interaction is used to give an estimate of the uncertainty in the calculation. Again the quoted error makes no allowance for the correctness of the model.

In table 1, we list the δ_{C2} values obtained by Towner, Hardy & Harvey[10] and Ormand & Brown[4] and note there is very good agreement between the two, except for the lightest case ^{14}O. Here there is a worrisome discrepancy with Ormand & Brown estimating a large correction (larger than any of the other seven cases). This is a little difficult to understand since the mismatch between the proton and neutron separation energies, the principal ingredient in δ_{C2}, decreases with lighter nuclei, so δ_{C2} perhaps should decrease in lighter nuclei.

3.2. The Correction, δ_{C1}

The other contribution to δ_C comes from isospin mixing within an isobaric multiplet. This can be estimated in a shell-model calculation. Towner & Hardy[11] first did this using a charge-dependent residual interaction by adjusting a selected nucleon-nucleon interaction in the

Table 1. The Coulomb correction, δ_c, from isospin mixing, δ_{c1}, and mismatch in radial overlap integrals, δ_{c2}.

	δ_{c1} (%)		δ_{c2} (%)		δ_c (%)
	TH[a]	OB[b]	THH[c]	OB[b]	
^{14}O	0.00	0.01	0.28±0.03	0.68±0.14	0.30±0.08
^{26}Al	0.06	0.01	0.27±0.04	0.43±0.05	0.37±0.08
^{34}Cl	0.02	0.06	0.62±0.07	0.66±0.07	0.68±0.05
^{38}K	0.16	0.11	0.54±0.07	0.46±0.07	0.64±0.06
^{42}Sc	0.04	0.11	0.35±0.06	0.46±0.07	0.47±0.07
^{46}V	0.09	0.01	0.36±0.06	0.41±0.07	0.43±0.06
^{50}Mn	0.10	0.00	0.40±0.09	0.50±0.06	0.52±0.07
^{54}Co	0.03	0.01	0.56±0.06	0.54±0.05	0.56±0.05

a) ref. 12; b) ref. 4; c) ref. 10

following way: *(i)* add two-body Coulomb terms in the proton-proton part of the Hamiltonian, *(ii)* increase the T=1 part of the proton-neutron Hamiltonian by 2% (justified by the charge dependence observed in nucleon-nucleon scattering data), and *(iii)* determine the one-body part of the Hamiltonian from the single-particle energies from closed-core plus proton and neutron nuclei.

Ormand & Brown[3,4] suggested an alternative strategy for determining the charge-dependent interaction empirically by requiring that the parameters of a Coulomb plus phenomenological isovector and isotensor potential reproduce experimental b- and c-coefficients of the isobaric-mass-multiplet equation. They argued that this procedure better determined the single-particle energies for nuclei away from a closed major shell. Towner & Hardy[12] have repeated their calculation adopting a variation on this strategy. They insisted that the b- and c-coefficients for the isobaric multiplet involved in the Fermi beta decay be fitted in the procedure. The result of their calculation together with that of Ormand & Brown is given in table 1. The results are sensitive to the choice of the shell model space and the underlying nucleon-nucleon interaction. The values are generally small, $\delta_{c1} \lesssim 0.1\%$, with the scatter between the two calculations being typically less than the uncertainty in the other part of the correction, δ_{c2}.

3.3. The Summed Correction

We combine the corrections δ_{C1} and δ_{C2} in the following way to arrive at a recommended δ_C value: (i) take a mean of the two δ_{C1} calculations and assign an error equal to the spread, (ii) take a weighted average of the two δ_{C2} calculations and assign an error equal to one standard deviation, and (iii) sum the δ_{C1} and δ_{C2} values and add the errors in quadrature. The results are in table 1. It is clear that it is difficult to reduce the uncertainty in the δ_C calculation below 0.05%. Ultimately this uncertainty will limit the tests of weak interaction physics available from data on superallowed β-decay.

4. EFFECTIVE VECTOR COUPLING CONSTANT

The test of the conserved vector current (CVC) hypothesis is that the ft values for superallowed β-decay, when suitably corrected for electromagnetic effects as just discussed, should all be constant. We write

$$ft(1+\delta_R)(1-\delta_C) \equiv \mathscr{F}t = K/(2G_v'^2)$$

$$K = 2\pi^3 \, \ell n2 \, \hbar^7/m^5 c^4$$

$$G_v'^2 = G_v^2 (1 + \Delta_\beta) \tag{3}$$

where G_v' is the effective vector coupling constant. It includes the nucleus-independent inner radiative correction, Δ_β, which is not relevant for the test of CVC. The status of the experimental data on the eight accurately measured superallowed transitions is under review[12]. Preliminary values for the experimental ft-values, together with corrections δ_R and δ_C are given in table 2. The corrected $\mathscr{F}t$-values for the eight cases are mutually consistent and the average value is

$$\mathscr{F}t = 3070.89 \pm 1.14 \text{ s} \tag{4}$$

The χ^2 per degree of freedom is one. The recommended effective coupling constant is

$$G_v'/(\hbar c)^3 = (1.14982 \pm 0.00021) \times 10^{-5} \text{ GeV}^{-2} \tag{5}$$

Table 2: ft-values, corrections δ_R and δ_C, and the corrected $\mathcal{F}$t-values for the eight accurately measured superallowed Fermi β-transitions.

	ft	δ_R(%)	δ_C(%)	$\mathcal{F}$t
^{14}O	3037.1±1.6	1.52±0.01	0.30±0.08	3073.9±3.0
^{26}Aℓ	3033.8±1.2	1.46±0.02	0.37±0.08	3066.6±2.8
^{34}Cℓ	3050.0±1.9	1.42±0.03	0.68±0.05	3072.4±2.7
^{38}K	3047.8±2.7	1.42±0.04	0.64±0.06	3071.3±3.4
^{42}Sc	3040.9±2.0	1.44±0.05	0.47±0.07	3070.1±3.2
^{46}V	3043.2±2.2	1.43±0.06	0.43±0.06	3073.3±3.3
^{50}Mn	3039.5±4.0	1.43±0.07	0.52±0.07	3066.9±4.9
^{54}Co	3044.8±2.3	1.42±0.07	0.56±0.05	3070.7±3.6

5. J≠0 SUPERALLOWED BETA DECAYS

To deduce the weak vector coupling constant from J≠0 superallowed beta decays requires not only an accurate lifetime measurement but also a correlation experiment so that the Gamow-Teller (axial-vector) fraction can be separated from the Fermi (vector) fraction. The generalisation of eq. (3) is

$$ft(1+\delta_R)(1-\delta_C) \equiv \mathcal{F}t = K/[G'^2_V \langle M_F \rangle^2 (1+\rho^2)] \tag{6}$$

where $\langle M_F \rangle$ is the Fermi matrix element ($\sqrt{2}$ for T=1 states) and ρ is the ratio $(g_A/g_V)\langle M_{GT}\rangle/\langle M_F\rangle$ where $\langle M_{GT}\rangle$ is the Gamow-Teller matrix element and g_A/g_V the ratio of axial-vector to vector coupling constants. (The phase space integral f is also slightly different between vector and axial-vector transitions and this leads to an additional small correction to eq. (6)). A correlation experiment is needed to determine ρ. This has been achieved in five cases. For the neutron (ref. 13), ^{19}Ne (ref. 14) and ^{35}Ar (ref. 15) the polarised-nucleus electron-direction correlation is measured; for ^{20}Na (ref. 16) the beta-neutrino-alpha triple-correlation is obtained; while for ^{24}Aℓ (ref. 17) the value of ρ is so small that a theoretical calculation will suffice. The result for the weak vector coupling in each case is:

neutron: $G'_V/(\hbar c)^3 = (1.147 \pm 0.007) \times 10^{-5} \text{ GeV}^{-2}$

^{19}Ne: $G'_V/(\hbar c)^3 = (1.152 \pm 0.002) \times 10^{-5} \text{ GeV}^{-2}$

^{20}Na: $G'_V/(\hbar c)^3 = (1.103 \pm 0.028) \times 10^{-5} \text{ GeV}^{-2}$

^{24}Aℓ: $G'_V/(\hbar c)^3 = (1.148 \pm 0.008) \times 10^{-5} \text{ GeV}^{-2}$

^{35}Ar: $G'_V/(\hbar c)^3 = (1.136 \pm 0.008) \times 10^{-5} \text{ GeV}^{-2}$

The weighted average is

$$G'_V/(\hbar c)^3 = (1.1511 \pm 0.0019) \times 10^{-5} \text{ GeV}^{-2} \tag{7}$$

which is in excellent accord with the more precise value obtained from the average of eight $J = 0^+$ superallowed emitters, eq. (5). The consistency between the $J \neq 0$ and $J = 0$ results indicates there is no evidence for any spin-dependent effects in the weak vector coupling constant.

6. MUON DECAY COUPLING CONSTANT

A comparison of semi-leptonic weak decays, such as the superallowed β-decays, with a pure leptonic decay, such as muon decay leads to a value of V_{ud}, the leading element in the Kobayashi-Maskawa mixing matrix to be discussed below. The muon decay mean lifetime is related to the muon coupling constant by the expression[18,19]

$$\frac{\hbar}{\tau_\mu} = \frac{1}{192\pi^3} \frac{G_\mu^2}{(\hbar c)^6} (m_\mu c^2)^5 f_1 [1 + \delta^{(2)} + \Delta_\mu] \tag{8}$$

where f_1 is a recoil correction ($f_1 = 0.99813$) an $\delta^{(2)}$ the outer radiative correction of order α ($\delta^{(2)} = -0.42\%$). An additional term, Δ_μ, is a radiative correction arising from Z-boson exchanges between the muon and electron, which, in analogy with nuclear β decay, is incorporated into an effective coupling constant: $G'^2_\mu = G^2_\mu(1+\Delta_\mu)$. The muon mean lifetime from the 1986 Particle Data Group review[20] is $\tau_\mu = (2.197033 \pm 0.000039) \times 10^{-6}$ s. Hence

$$G'_\mu/(\hbar c)^3 = (1.166352 \pm 0.000013) \times 10^{-5} \text{ GeV}^{-2} \tag{9}$$

and the ratio of vector to muon coupling constants is

$$G_v'^2/G_\mu'^2 = V_{ud}^2 (1 + \Delta_\beta - \Delta_\mu)$$

$$= 0.9719 \pm 0.0004 \qquad (10)$$

7. INNER RADIATIVE CORRECTIONS

To derive V_{ud} an estimate of the inner radiative correction is needed. Note that only the difference $\Delta_\beta - \Delta_\mu$ is required so that universal contributions to both β-decay and μ-decay do not have to be calculated. In the simplest version of the Weinberg-Salam model of the electroweak interactions, Sirlin[18] derives

$$\Delta_\beta - \Delta_\mu = \frac{\alpha}{2\pi} [4\ell n(m_z/m_P) + \ell n(m_P/m_A) + 2C + \mathcal{A}_g + \cdots] \qquad (11)$$

where m_P, m_z, m_A are the masses of the proton, Z-vector boson, and A_1-meson respectively. In eq. (11) the term $\mathcal{A}_g$ is a small perturbative QCD correction that Marciano & Sirlin[21] have estimated to be -0.34. The second term, $\ell n(m_P/m_A) + 2C$, is the most troublesome piece and is the only contribution of order α which is not fully under control. It is a structure-dependent term that arises from the axial-vector current; the mass m_A represents a low-energy cut-off for this piece of the calculation. Marciano and Sirlin suggest a generous range: 400 MeV $\leq$ m_A $\leq$ 1600 MeV that spans the physical mass of the A_1-meson. The constant, C, is model dependent but a range of $0 \leq C \leq 1$ spans the estimates[21,22] given so far. Allowing for the above ranges, and using a systematic error, Marciano & Sirlin[21] recommend

$$\frac{\alpha}{2\pi} [\ell n(m_P/m_A) + 2C] = 0.0012 \pm 0.0018 \qquad (12)$$

The largest term in eq. (11), $4\ell n(m_z/m_P)$, is unambiguous. Marciano & Sirlin have, in addition, estimated the effect of higher orders by summing all leading-logarithmic corrections to order $\alpha^n \ell n^n m_z$, with $n = 1, 2, \cdots$ via a renormalisation-group analysis. The result is an estimate of the inner radiative correction of

$$\Delta_\beta - \Delta_\mu = (2.37 \pm 0.18) \times 10^{-2} \qquad (13)$$

where the error is entirely due to the uncertainty given in eq. (12). Combining this with eq. (10) gives

$$|V_{ud}| = 0.9743 \pm 0.0009 \qquad (14)$$

for the leading element in the Kobayashi-Maskawa mixing matrix. The error is dominated by the uncertainty in the inner radiative correction.

8. KOBAYASHI-MASKAWA MIXING MATRIX

In the standard model of electroweak interactions, both quarks and leptons are assigned to be left-handed doublets and right-handed singlets. The quark mass eigenstates are not the same as the weak eigenstates and the matrix connecting them has become known as the Kobayashi-Maskawa mixing matrix[23]. By convention the three 2/3 charge quarks (u, c and t) are unmixed and all the mixing is expressed in terms of a 3x3 unitary matrix operating on the charge -1/3 quarks (d,s,b):

$$\begin{pmatrix} d' \\ s' \\ b' \end{pmatrix} = \begin{pmatrix} V_{ud} & V_{us} & V_{ub} \\ V_{cd} & V_{cs} & V_{cb} \\ V_{td} & V_{ts} & V_{tb} \end{pmatrix} \begin{pmatrix} d \\ s \\ b \end{pmatrix}$$

In general, a 3x3 unitary matrix depends on four parameters so there are many inter-relations amongst the nine elements of the matrix. In particular the first row provides a unitarity test

$$|V_{ud}|^2 + |V_{us}|^2 + |V_{ub}|^2 = 1 \qquad (15)$$

which should be satisfied if the standard model of three generations is correct.

The matrix element V_{us} is determined from kaon and hyperon decays. In principle the data from kaon decays are to be preferred because these are vector transitions and corrections to an analysis based on exact SU(3) wavefunctions (in which quark masses are degenerate, $m_u = m_d = m_s$) are of second order in the symmetry breaking. In the past, there has been a problem in that the charge K^+_{e3} and neutral K^0_{e3} decays were incompatible. An analysis by Leutwyler & Roos[24] including isospin mixing ($m_u \neq m_d$) has now resolved this difficulty bringing the values for $|V_{us}|$ extracted from these two decays into agreement at the 1% level of accuracy. Their recommended value is $|V_{us}|$ = 0.2196 ± 0.0023.

The hyperon decays, on the other hand, involve both vector and axial-vector transitions and symmetry breaking effects occur in first order. Despite that, fits to the hyperon data assuming exact SU(3) symmetry are excellent; a typical result[25] based on the CERN hyperon data is $|V_{us}|$ = 0.231 ± 0.003. Nevertheless, symmetry breaking is known to be important. In other contexts, quark-model calculations have been shown to provide a good guide to the size of first-order SU(3) breaking caused by the s and u,d quark mass difference. Unfortunately the use of the quark-model symmetry breaking did not lead to an improved fit to the hyperon data; in many cases the fit was worse. Recently Donoghue, Holstein & Klimt[26] realised that earlier calculations omitted a crucial ingredient -- centre-of-mass or recoil corrections. With this ingredient they find the quark-model pattern of symmetry breaking leads to a significantly improved description of hyperon decays. Their final value is $|V_{us}|$ = 0.220 ± 0.001 ± 0.003 where the first error is experimental and the second an estimate of theoretical uncertainty. Thus kaon and hyperon data are now in agreement. The recommended value for V_{us} is

$$|V_{us}| = 0.220 \pm 0.002 \tag{16}$$

Until recently, there has only been a bound on V_{ub} based on the b-quark lifetime and bounds on the branching ratio, $\Gamma(b \to u)/\Gamma(b \to c)$, yielding $|V_{ub}|$ < 0.008. A new experiment, reported at the 1988 Rockport Conference[27], presented evidence for the observation of the rare decay $B^+ \to \bar{p}p\pi$, that would indicate a non-zero value for V_{ub}. The analysis involves some model-dependent assumptions. The result is

$$|V_{ub}| = 0.0055 \pm 0.0025 \tag{17}$$

Using eqs. (14), (16) and (17) the unitarity test, eq. (15), becomes

$$|V_{ud}|^2 + |V_{us}|^2 + |V_{ub}|^2 = 0.998 \pm 0.002 \tag{18}$$

consistent with a three-generation standard model. This result is considered a triumph[21] for the calculation of radiative corrections.

9. PION β-DECAY

There is one other superallowed β-decay at our disposal and this is the decay: $\pi^+ \to \pi^0 + e^+ + \nu_e$. It proceeds solely through the weak vector interaction. The branching ratio for this decay mode is small, 10^{-8}. The partial mean lifetime is given by the expression[18]

$$\frac{\hbar}{\tau} = \frac{1}{30\pi^3} \frac{G_V^2}{(\hbar c)^6} \Delta^5 f_1 f_2 (1 + \delta_R^\pi + \Delta_\pi)$$ (19)

where $\Delta = m_+ - m_0$ is the pion mass difference, $\Delta = 4.5930 \pm 0.0013$ MeV (ref. 28). Here Δ^5 is the leading term in the phase space integral, f_1 its correction factor ($f_1 = 0.941023$), f_2 is a recoil correction ($f_2 = 0.951445$), and δ_R^π and Δ_π are outer and inner radiative corrections. If in this expression we replace G_V^2 by its value from superallowed β-decay, $G_V^2 = (K/2\mathscr{F}t)(1-\Delta_\beta)$, then

$$\tau^{-1} = \frac{\ell n2}{\mathscr{F}t} \frac{1}{30} \left(\frac{\Delta}{m_e c^2} \right)^5 f_1 f_2 (1 + \delta_R^\pi + \Delta_\pi - \Delta_\beta)$$

$$= (0.3952 \pm 0.0006) (1 + \delta_R^\pi + \Delta_\pi - \Delta_\beta) \ s^{-1}$$ (20)

where the error is entirely due to the uncertainty in Δ, the pion mass difference. The outer radiative correction to order α, δ_R^π, is given by the same expression[18] as used in nuclear β-decay and is $(1.05 \pm 0.15)\%$. The inner radiative corrections for pion and nuclear β-decay are expected to be the same, $\Delta_\pi - \Delta_\beta \simeq 0$; hence

$$\tau^{-1} = 0.3993 \pm 0.0008 \ s^{-1}$$ (21)

Two recent experiments from LAMPF give $\tau^{-1} = 0.398 \pm 0.015 \ s^{-1}$ (ref. 29) and $\tau^{-1} = 0.394 \pm 0.015 \ s^{-1}$ (ref. 30) in excellent agreement with theory.

10. NEUTRON DECAY

Finally, we would like to combine our value of $\mathcal{F}t$ from superallowed β-decay with the recently determined value of $g_A/g_V = 1.261 \pm 0.004$ (from ref. 13 and averaged with earlier data) to deduce a value for the neutron mean lifetime, τ_n. The connection between these quantities is

$$\tau_n = \frac{2\,\mathcal{F}t}{f(1+\delta_R)} \cdot \frac{1}{\ell n2}\,[1 + 3(g_A/g_V)^2]^{-1}$$

$$= 895.6 \pm 4.7\ \text{s} \tag{22}$$

where the phase space integral and radiative correction are taken from Wilkinson[31]: $f(1+\delta_R) = 1.71465 \pm 0.00015$ and the $\mathcal{F}t$ value is from eq. (4). Until recently there were three measurements of the neutron mean lifetime that unfortunately were incompatible: 918 ± 14 s (ref. 32), 881 ± 8 s (ref. 33) and 937 ± 18 s (ref. 34). There has been a temptation in the past (e.g. ref. 25) to average the two high values and reject the third value. This produces $\tau_n = 925 \pm 11$ s in disagreement with eq. (22). An alternative strategy and one we favour is to accept all three lifetime measurements and scale the errors so that χ^2 on the average is one. This produces $\tau_n = 898 \pm 16$ s. Two new measurements have recently been published that vindicate this second approach. They are 903.0 ± 13.0 (ref. 35) and 876.0 ± 21.0 (ref. 36). When the five data are averaged together and the error scaled, the resulting best value is

$$\tau_n = 895.6 \pm 9.8\ \text{s} \tag{23}$$

in excellent agreement with the value deduced from the superallowed $\mathcal{F}t$-value and correlation data, eq. (22).

11. CONCLUSION

In summary, we note:

(a) The 1984 noted discrepancy in the $\mathcal{F}t$ values of five standard deviations between low-Z and high-Z superallowed β-decay data has been resolved by the new radiative correction calculations in order $Z\alpha^2$ of Sirlin et al.[2].

225

(b) The principal uncertainty in the deduced $\mathcal{F}t$ values seems to be the nuclear structure correction, δ_c. The large difference between the Ormand-Brown[4] and Towner-Hardy-Harvey[10] calculations for ^{14}O is worrisome. A precision measurement on ^{10}C decay could shed some light on this problem.

(c) The unitarity test on the first row of the Kobayashi-Maskawa matrix is satisfied for three generations of quarks in the standard model. The principal uncertainty in the test is the theoretical uncertainty in the inner radiative correction.

(d) Experiments on pion β decay are consistent with the predictions from nuclear β decay. An order of magnitude improvement in the measurement of the branching ratio in pion decay could in principle provide a test of the nuclear structure corrections in nuclear β decay.

(e) The neutron lifetime determined from the $\mathcal{F}t$-value in superallowed β-decay and the measurement of g_A/g_V in a correlation experiment is in agreement with, but more accurate than, direct lifetime measurements achieved so far.

References

1. Towner, I.S. and Hardy, J.C., in Proc. 7th Int. Conf. on Atomic Masses and Fundamental Constants, Darmstadt-Seeheim 1984, ed. O. Klepper (Technische Hochschule, Darmstadt) p.564
2. Sirlin, A. and Zucchini, R., Phys. Rev. Lett. 57, 1994 (1986)
3. Ormand, W.E. and Brown, B.A., Nucl. Phys. A440, 274 (1985)
4. Ormand, W.E. and Brown, B.A., preprint 1988, and private communication
5. Sirlin, A., Nucl. Phys. B71, 29 (1974)
6. Jaus, W. and Rasche, G., Nucl. Phys. A143, 202 (1970); Jaus, W., Phys. Lett. 40B, 616 (1972)
7. Jaus, W. and Rasche, G., Phys. Rev. D35, 3420 (1987)
8. Sirlin, A., Phys. Rev. D35, 3423 (1987)
9. Wilkinson, D.H., Phys. Lett. 65B, 9 (1976)
10. Towner, I.S. Hardy, J.C. and Harvey, M., Nucl. Phys. A284, 269 (1977)

226

11. Towner, I.S. and Hardy, J.C., Nucl. Phys. A205, 33 (1973)

12. Towner, I.S. and Hardy, J.C., to be published

13. Klemt, E. et al., Z. Phys. C37, 179 (1988)

14. Calaprice, F.P. et al., Phys. Rev. Lett. 35, 1566 (1975)

15. Garnett, J.D. et al., Phys. Rev. Lett. 60, 499 (1988)

16. Clifford, E.T.H. et al., Phys. Rev. Lett. 50, 23 (1983); and to be published.

17. Adelberger, E.G. et al., Phys. Rev. Lett. 55, 2129 (1985)

18. Sirlin, A., Rev. Mod. Phys. 50, 573 (1978)

19. Compilation of coupling constants and low-energy parameters, Nucl. Phys. B216, 277 (1983)

20. Review of particle properties, Phys. Lett. 170B, 1 (1986)

21. Marciano, W.J. and Sirlin, A., Phys. Rev. Lett. 56, 22 (1986)

22. Abers, E.S. et al., Phys. Rev. 167, 1461 (1968);
    Dicus, D. and Norton, R., Phys. Rev. D1, 1360 (1970);
    Sirlin, A., Nucl. Phys. B71, 29 (1974)

23. Kobayashi, M. and Maskawa, K., Prog. Theo. Phys. 49, 652 (1983)

24. Leutwyler, H. and Roos, M., Z. Phys. C25, 91 (1984)

25. Bourquin, M. et al., Z. Phys. C21, 27 (1983)

26. Donoghue, J.F., Holstein, B.R. and Klimt, S.W., Phys. Rev. D35, 934 (1987)

27. Schubert, K., 3rd Conf. on the Intersections between particle and nuclear physics, Rockport 1988, to be published

28. Crawford, J.F. et al., Phys. Rev. Lett. 56, 1043 (1986)

29. McFarlane, W.K. et al., Phys. Rev. Lett. 51, 249 (1983)

30. McFarlane, W.K. et al., Phys. Rev. D32, 547 (1985)

31. Wilkinson, D.H., Nucl. Phys. A377, 474 (1982)

32. Christensen, C.J. et al., Phys. Rev. D5, 1628 (1972)

33. Bondarenko, L.N. et al., JETP Lett. 28, 303 (1978)

34. Byrne, J. et al., Phys. Lett. 92B, 274 (1980)

35. Kosvintsev, Yu. Yu., Morozov, V.I. and Terekhov, G.I., JETP Lett. 44, 571 (1986)

36. Last, J. et al., Phys. Rev. Lett. 60, 995 (1988)

WHY IS THE ELECTRON SO LIGHT
AND WHY IS THE NEUTRINO SO MUCH LIGHTER?*

Aharon Davidson

McGill University

3600 University Street, Montreal,Quebec, Canada H3A 2T8.†

ABSTRACT

The Universal Seesaw mechanism, designed for $m_{e,u,d} \ll m_W$, serendipitously predicts $m_{\nu_L} m_{\nu_R} = m_e^2$. Extended to higher generations, the scheme exhibits a family mass hierarchy, but does not allow for an analogous hierarchy among the automatically superlight neutrinos. Neutrino masses are governed by axion physics: $m_\nu = 10\ eV \leftrightarrow m_{PQ} = 10^{12}$ GeV via $m_\nu m_{PQ} = m_W^2$, and L-R symmetry is spontaneously violated. Alternatively, $m_\nu = 10^{-2} eV \leftrightarrow m_{PQ} = 10^{10}$ GeV via $m_\nu m_{PQ} = m_e m_W$, and L-R symmetry is explicitly broken. The group theoretical consistency of the seesaw fermions is demonstrated by means of a hybrid unification scheme.

The origin of quark and lepton masses is well established within the framework of the standard electro/weak theory. However, reflecting the complete arbitrariness of the Yukawa sector, this theory lacks the ability to account for their actual masses. Of exceptional theoretical frustration are the mass hierarchies.

$$m_e \simeq 10^{-5} m_W \ ,$$

$$m_\nu \overset{<}{\sim} 10^{-5} m_e \ .$$
(1)

The superlightness of the neutrino is conventionally attributed to the so-called 'seesaw mechanism,' specifically designed to allow for

$$m_{\nu_L} m_{\nu_R} \simeq m_e^2$$
(2)

---

*Talk given at the Summer Institute in Theoretical Physics (Queen's University, Kingston, Ontario, Canada, July 1988).

† Permanent address: Physics Department, Ben-Gurion University of the Negev, Beer-Sheva 84105, Israel.

It may well be that a universal generalization of the seesaw idea is responsible for the lightness of $m_{e,u,d}$ on the m_W-scale. In which case, the above hierarchies better be correlated. A minimal $SU(3)_C \times SU(2)_L \times SU(2)_R \times U(1)_{B-L}$ realization is hereby discussed.

The standard **complex** fermionic representation

$$q_L(3,2,1)_{1/3} + q_R(3,1,2)_{1/3}$$

$$\ell_L(1,2,1)_{-1} + \ell_R(1,1,2)_1 \quad , \tag{3}$$

is supplemented by the **real** piece.

$$U_{L,R}(3,1,1)_{4/3} + D_{L,R}(3,1,1)_{-2/3} \quad , $$

$$N_{L,R}(1,1,1)_{\circ} + E_{L,R}(1,1,1)_{-2}. \tag{4}$$

The idea is that every ordinary fermion has an $SU(2)_L \times SU(2)_R$ – singlet companion with matching $SU(3)_C \times U(1)_Q$ assignments. While enlarging the fermionic representation, the Higgs system is simplified to its limits. The fermions Yukawa couple to

$$\phi_L(1,2,1)_{-1} + \phi_R(1,1,2)_{-1}. \tag{5}$$

Especially notice that the conventional sources of quark and lepton masses, namely $\Phi(1,2,2)_{\circ} + \Phi(1,3,1)_{-2} + \Phi(1,1,3)_{-2}$, have **not** been introduced. A parity-odd scalar $\sigma(1,1,1)_{\circ}$ is optional; it is mandatory if the L-R symmetry is to be spontaneously violated.

Now, let $L, R \equiv <\phi_{L,R}>$, and let Λ denote the bare mass of seesaw fermions. The generic single-generation quark mass matrices then look like

$$M_u \sim \begin{pmatrix} 0 & L \\ R^* & \Lambda \end{pmatrix}, M_d \sim \begin{pmatrix} 0 & L^* \\ R & \Lambda \end{pmatrix}, \tag{6}$$

up to Yukawa couplings of the same order of magnitude. With $L \ll R \ll \Lambda$, it is trivial to verify that

$$m_{e,u,d} \sim \frac{LR}{\Lambda} \ll m_W. \tag{7}$$

At the low-energy limit ($\Lambda \to \infty$), the above result is described by a dim -5 operator $\frac{1}{\Lambda} \left(\phi_L^{\dagger} \phi_R \right) \overline{\psi_L} \psi_R$. In other words, the old scalar $\phi(1,2,2)_{\circ}$ is represented by $\phi_L^{\dagger} \phi_R$. One may further observe that

$$\phi(1,3,1)_{-2} \equiv \phi_L^2 \quad , \quad \phi(1,1,3)_{-2} \equiv \phi_R^2 \quad , \tag{8}$$

and consequently expect that, with $L \ll R \ll \Lambda$ already established to make $m_e \ll m_W$, the superlightness of the neutrino will follow at no extra cost. Indeed, still suppressing the Yukawa couplings, the neutrino mass matrix

$$M_\nu \sim \begin{pmatrix} 0 & 0 & L & L \\ 0 & 0 & R & R \\ L & R^* & \Lambda & \Lambda \\ L & R^* & \Lambda & \Lambda \end{pmatrix} \tag{9}$$

tells us that the lowest eigenmasses are

$$m_{\nu_L} \sim \frac{L^2}{\Lambda} \quad , \quad m_{\nu_R} \sim \frac{R^2}{\Lambda}. \tag{10}$$

This is the neutrino bonus: $m_e \ll m_W$ serendipitously predicts $m_{\nu_L} m_{\nu_R} \simeq m_e^2$.

Given $\frac{m_\nu}{m_e} \sim \frac{L}{R}$ and $\frac{m_e}{m_W} \sim \frac{R}{\Lambda}$, let us numerically discuss the various mass scales involved. The simple-minded case is

$$m_\nu = 10eV \Rightarrow R \sim 10^7 \text{ GeV} \quad , \quad \Lambda \sim 10^{12} \text{ GeV}. \tag{11a}$$

This may be a hint that the neutrino mass is governed by axion physics. The corresponding signature is relatively light $\frac{R^2}{\Lambda} \sim 10^2$ GeV right-handed neutrinos. In fact, as we shall see, the multi-generational generalization of the universal seesaw idea does provide a compelling reason for $m_\nu \sim \frac{1}{m_{PQ}}$.

The other alternative

$$m_\nu = 10^{-2} \ eV \Rightarrow R \sim 10^{10} \text{ GeV} \quad , \quad \Lambda \sim 10^{15} \text{ GeV} \tag{11b}$$

is potentially relevant for the solar neutrino problem. Here we face the GUT scale, and correspondingly $\frac{R^2}{\Lambda} \sim 10^5$ GeV.

However, as it stands, the scheme is incomplete. The standard question why is the electron so light has been replaced by the question why is the (say) top quark so heavy. With regard to the flavor puzzle we now argue that without appealing to a hierarchy in the Yukawa sector, and without introducing new mass scales, quarks and leptons may come in two different mass scales. All one needs is a simple **axial** $U(1)_A$ symmetry principle. The point is quite tricky. If the Λ-term is absent, on some symmetry grounds, (6) is replaced by

$$M \sim \begin{pmatrix} 0 & L \\ R & 0 \end{pmatrix}, \tag{12}$$

which in turn implies

$$m_f \sim L \sim m_W. \tag{13}$$

Thus, to construct a tenable multigenerational scheme, a **singular** sub-matrix Λ is required. But this cannot be enforced by means of a vectorial $U(1)_V$ (the diagonal Λ entries stay unrestricted). In other words, **a singular Λ calls for a horizontal $U(\Lambda)_A$**. This conclusion is favored by grand unification and especially by the strong CP problem.

Amusingly, once an extra symmetry is invoked to prevent some fermionic masses of order L from dropping into the $\frac{LR}{\Lambda}$ regime, the traditional role of a symmetry has apparently turned upside down. However, this is to be regarded as a universal seesaw artifact.

At any rate, a specific model is in order. At the two-generation level, the unique example has to do with

$$M \sim \begin{pmatrix} 0 & 0 & 0 & L \\ 0 & 0 & L' & \ell \\ 0 & R' & \Lambda & 0 \\ R & r & 0 & 0 \end{pmatrix}. \qquad (14)$$

The spectrum consists of two ordinary families of masses

$$m_1 \sim \frac{LR}{\Lambda} \quad , \quad m_2 \sim L \quad , \qquad (15)$$

and their superheavy seesaw partners $m_1' \sim \Lambda$, $m_2' \sim R$. The corresponding axial charges are $Q(f_{L,R}^1) = \pm 3$, $Q(f_{L,R}^2) = \pm 1$, $Q(F_{L,R}^1) = 0$, and $Q(F_{L,R}^2) = \mp 2$, conveniently normalized such that the scalars carry $Q(\phi) = \pm 1$.

With the $U(1)_A$ assignments already established to account for the charged fermion mass hierarchy, do we expect an analogous hierarchy in the neutrino sector? To answer this question, one simply doubles the structure of Eq. (14) to form an 8 x 8 matrix, and extracts the lowest eigen masses. The result

$$m_{\nu_1} \sim m_{\nu_2} \sim \frac{L^2}{\Lambda} \qquad (16)$$

is a big surprise: **No mass hierarchy is allowed among the automatically superlight left-handed neutrinos.**

$U(1)_A$ is furthermore advantageous in the sense that one may use the cosmological window 10^9 GeV $< m_{PQ} < 10^{12}$ GeV to probe neutrino masses. There are two options:

1. In the absence of the $\sigma(1, 1, 1,)_\circ$ scalar, with L-R symmetry explicitly broken, we have $m_{PQ} \sim R$. Consequently,

$$m_\nu \sim \frac{m_W m_e}{m_{PQ}} \qquad (17a)$$

is favored by 10^{-2} eV neutrinos ($m_{PQ} = 10^{10}$ GeV), but cannot support heavier neutrinos.

2. If $\sigma(1,1,1)_0$ is present, and L-R symmetry is spontaneously violated, it is mandatory that $m_{PQ} \sim \Lambda$. In turn,

$$m_\nu \sim \frac{m_W^2}{m_{PQ}} \qquad (17b)$$

establishes a fundamental correlation

$$m_\nu = 10 \ eV \leftrightarrow m_{PQ} \sim 10^{12} \text{ GeV} . \qquad (18)$$

Notice how restrictive this preferred alternative is: It is not compatible with lighter neutrinos!

Finally, to prove that the scheme is group theoretically consistent, we demonstrate its grand unifiability. The associated unified gauge group is $SU(5)_L \times SU(5)_R$, exhibiting Pati-Salam elements (discrete L-R symmetry, chiral color) and Georgi-Glashow ideas (chiral flavor). One ordinary generation plus its seesaw partner furnish the left-handed representation

$$\psi_L = (10 + 5^* + 1; 1) + (1; 10^* + 5 + 1). \qquad (19)$$

But one has to be extremely careful since $SU(3)_C \times SU(2)_L \times SU(2)_R \times U(1)_{B-L}$ is non-trivially embedded within $SU(5)_L \times SU(5)_R$. In fact, $SU(3)_C \equiv SU(3)_{L+R}$ and $U(1)_{B-L} \equiv U(1)_{L+R}$. It is worth noticing that ordinary left-handed (right-handed) fermions are $SU(5)_{R(L)}$-singlets. Moreover, $B - L$ is realized as the fourth-color without the parent $SU(4)_{PS}$ structure.

The novel feature has to do with the evolution of the weak mixing angle, reflecting the fact that $SU(3)_C \times SU(2)_L \times U(1)_Y$ does not fit into an overall $SU(5)$ factor. A simple-minded two-step spontaneous symmetry breaking gives rise to

$$\sin^2 \theta_W = \frac{3}{16} + \frac{\alpha}{3\pi} \left[N \ln \frac{M}{m_W} - \frac{143}{16} \ln \frac{M_G}{m_W} + \frac{55}{4} \ln \frac{M_G}{M} \right] \quad , \qquad (20)$$

where N =total number of generations, M =intermediate mass scale, and M_G =the unification mass. In particular notice that unlike in the Geogi Glashow $SU(5)$

1. $\sin^2 \theta_W$ starts from $\frac{3}{16}$ at the symmetry limit and rises, and

2. $\sin^2 \theta_W$ is N-dependent.

The experimental value of $\sin^2 \theta_W(M_W) \simeq 0.22$, and the fact that $N = 3,4$ imply an oasis at $M \sim 10^{7,8}$ GeV.

For further details see A. Davidson and K.C. Wali, Phys. Rev. Lett. **58**, 2623 (1987); Phys. Rev. Lett., **59**, 393 (1987); Phys. Rev. Lett. **60**, 1813 (1988).